图示水泥回转窑耐火砖砌筑

ILLUSTRATION OF CEMENT ROTARY KILN BRICK LINING

主 编 石 玉 张 超
Editor-in-Chief　SHI Yu　ZHANG Chao

中国建设科技出版社有限责任公司
China Construction Science and Technology Press Co., Ltd.

北 京

图书在版编目（CIP）数据

图示水泥回转窑耐火砖砌筑 / 石玉，张超主编 .
北京：中国建设科技出版社有限责任公司，2025.3.
ISBN 978-7-5160-4386-8

I. TU754.1

中国国家版本馆 CIP 数据核字第 2025CC5067 号

图示水泥回转窑耐火砖砌筑
TUSHI SHUINI HUIZHUANYAO NAIHUOZHUAN QIZHU
主 编 石 玉 张 超

出版发行：	中国建设科技出版社有限责任公司
地　　址：	北京市西城区白纸坊东街 2 号院 6 号楼
邮政编码：	100054
经　　销：	全国各地新华书店
印　　刷：	北京雁林吉兆印刷有限公司
开　　本：	850mm×1168mm　1/32
印　　张：	3.25
字　　数：	80 千字
版　　次：	2025 年 3 月第 1 版
印　　次：	2025 年 3 月第 1 次
定　　价：	**36.00 元**

本社网址：www.jskjcbs.com，微信公众号：zgjskjcbs
请选用正版图书，采购、销售盗版图书属违法行为
版权专有，盗版必究。本社法律顾问：北京天驰君泰律师事务所，张杰律师
举报信箱：zhangjie@tiantailaw.com　举报电话：（010）63567684
本书如有印装质量问题，由我社事业发展中心负责调换，联系电话：（010）63567692

PREFACE
序 言

　　耐火砖对水泥回转窑的稳定、可靠运行十分重要。笔者长期从事水泥回转窑耐火砖的施工监督工作，在工作过程中经常发现筑炉工会对书面形式的砌筑规范产生误解，便萌生用三维设计软件制作图示来清晰地表达水泥回转窑耐火砖施工要求和规范的想法，毕竟"一张图胜过千言万语"，力争做到"读过即会做"。

　　书中内容中如有不当之处，欢迎行业从业人员指出和斧正。

　　在本书编写过程中，同事张超先生提出了诸多修改意见，在此致谢。

　　Refractory brick plays a very important role in cement rotary kiln reliable running,author has engaged in cement rotary kiln brick installation many years. During work, often foundmisunderstanding of installation regulations in written form by bricklayers. This inspired to use 3D software to make illustrations to express how to make correct brick lining in cement rotary kiln clearly.As the saying goes, "a picture is worth a thousand words", and try to realize "can do after reading".

　　Welcome industry practitioners pointed out and correct the improper content inthis small booklet.

　　Herein, thanks my colleague Mr. Zhang Chao for his many suggestions for revisionduring the booklet compiling.

CONTENTS

目 录

安全
Safety ··· 01

准备工作
Preparation work ·· 02

砌筑方法
Lining methods ··· 04

大小头尺寸之差
Cold face and hot face ·· 13

火泥
Mortar ·· 14

非碱性砖区域，只能湿砌
None basic brick zone, only mortar lining ··· 18

砌筑开始
Start lining ··· 24

耐火砖铺底
Bottom lining ·· 26

骑在环向焊缝处的砖环
Brick ring on welding seam ··· 29

耐火砖面打火泥
Mortar applied on bricks ·· 30

砖环封闭时的要点
Key points when close brick ring ·· 37

锁缝钢板和自制锁缝钢板夹持器
Closure shim and self-made closure shim holder ······ 41

锁缝砖和锁缝钢板纵向布置
Arrangement of key bricks and closure shims longitudinally ······ 43

新旧砖交接
The new lining meets the old lining ······ 44

新旧砖交接时耐火砖的加工
Brick cutting when new lining meets old lining ······ 46

局部挖补耐火砖
Partial Repairs(Patching) ······ 49

窑筒体变形，无法干砌
Shell deformed, clench lining cannot proceed ······ 50

挡砖圈，窑口
Brick retaining ring, outlet ······ 52

窑变径处砌筑
Brick lining on the cone ······ 56

使用 DAT 砌砖机进行砌筑
Lining with DAT lining machine ······ 58

DAT 砌砖机砌筑
Lining with DAT machine ······ 65

液压千斤顶顶住砖环缺口
Use the hydraulic jack to tighten the brick ring ······ 70

最后一环的最后一块砖
The last ring and the last brick ······ 71

当撤出 DAT 砌砖机时的最终检查
Final check when the DAT is withdrawn from the kiln ······ 73

窑内浇注料施工
Castable installation of rotary kiln ······ 76

记录和报告
Record and report ······ 79

碱性耐火砖的储存
Storage of basic bricks ······ 80

碱性耐火砖的水化——未衬砌之前损坏
Hydration of basic bricks—damage before Installation ······ 81

过高椭圆度造成的耐火砖寿命减少
Short brick service life caused by overdue ovality ······ 83

窑筒体椭圆度测量的必备条件
Precondition to check kiln shell ovality ······ 85

定期测量轮带滑移量
Checking tyre creep (migration) regularly ······ 86

窑筒体铺薄不锈钢板防腐
Sacrifice layer (thin stainless-steel sheet) in case severe kiln shell corrosion ······ 87

点火升温以及停窑后再次点火升温开窑
Heating-up and restart kiln again after kiln shut down ······ 88

回转窑中的主导损耗因素
Primary wear mechanisms in cement rotary kilns ······ 90

水泥回转窑机械状况常识
The common knowledge of cement rotary kiln mechanical condition ······ 91

未来可期
The future is promising ······ 95

参考文献
Reference ······ 96

安全
Safety

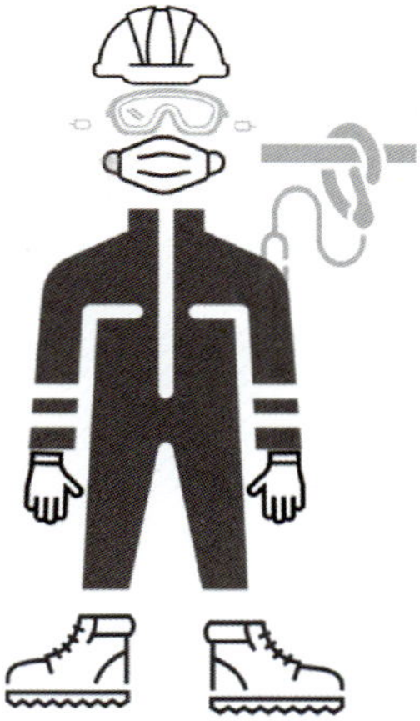	人身防护安全 Personnel safety protection
	作业场所和环境安全 Operating site & environment safety

准备工作
Preparation work

衬砌布置图

Lining plan

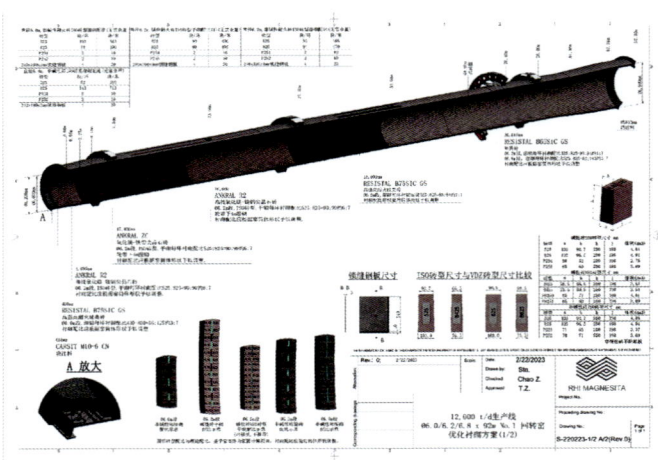

一般情况下，砌筑工作开展之前，工厂会做出一份砌筑布置文件，从砌筑布置可知：

— 衬砌厚度 — 每种耐火砖衬砌长度 — 砖型，配比，干砌还是湿砌 其他事项……	— 工期 — 安全设施 — 所需材料、砖、辅材 — 砌筑方法、设备及工具 — 人力 — 从仓库至窑头平台倒运 其他事项……

Usually, before start lining, the plant will prepares one lining plan document, from the lining plan:

- Lining thickness for the kiln - Each kind of brick lining length - Brick shape, mixing ratio, clench or mortar lining And other issues ...	- Schedule - Safety facilities - All material, brick, accessories - Lining method, equipments and tools - Manpower - Handling between warehouse and kiln platform And other issues ...

砌筑方法
Lining methods

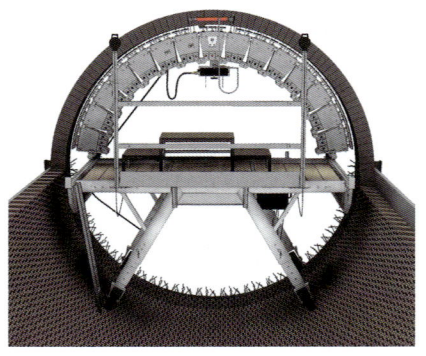

砌砖机法砌筑是水泥回转窑耐火砖砌筑的主流方法,是最安全和最高效的水泥回转窑耐火砖砌筑方法。

The mainstream method for lining cement rotary kilns is using a lining machine, which is the safest and most efficient method.

螺旋顶杠法砌筑
Screw Jacks method

螺栓槽钢法砌筑
Bolt (and channel) method

弹簧顶杆（Pogo sticks）法砌筑
Pogo sticks method

还有胶结法等不一而足。
And glue method etc.

螺旋顶杠方法
Screw Jacks method

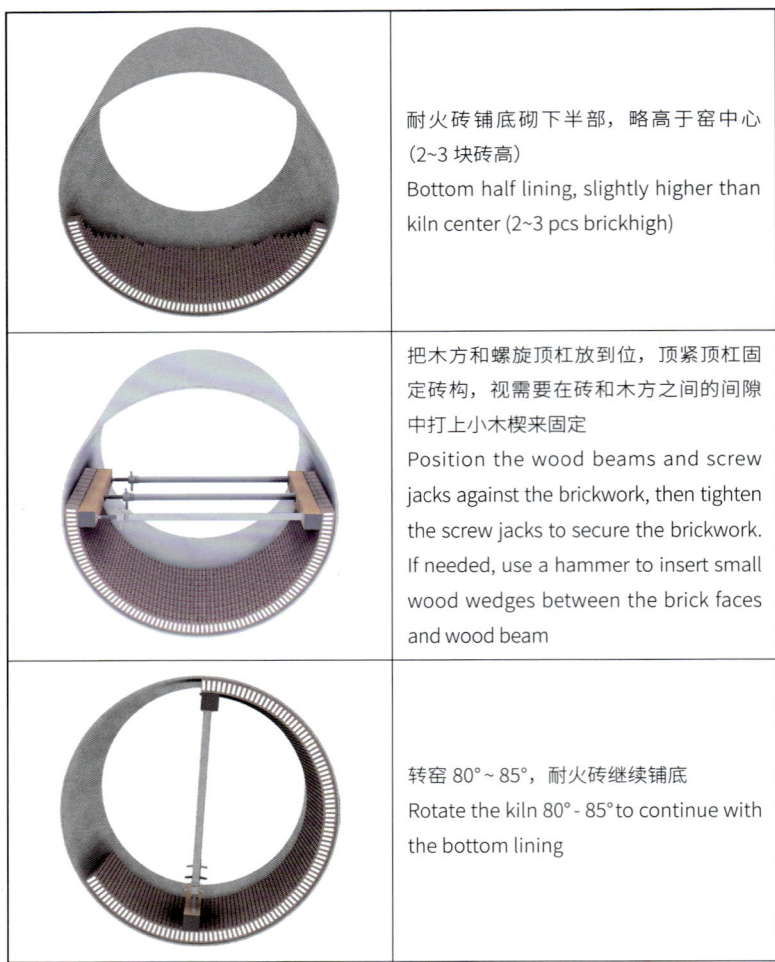

	耐火砖铺底砌下半部，略高于窑中心（2~3 块砖高） Bottom half lining, slightly higher than kiln center (2~3 pcs brickhigh)
	把木方和螺旋顶杠放到位，顶紧顶杠固定砖构，视需要在砖和木方之间的间隙中打上小木楔来固定 Position the wood beams and screw jacks against the brickwork, then tighten the screw jacks to secure the brickwork. If needed, use a hammer to insert small wood wedges between the brick faces and wood beam
	转窑 80°~ 85°，耐火砖继续铺底 Rotate the kiln 80°- 85°to continue with the bottom lining

续表

	继续耐火砖铺底，耐火砖铺到略高于窑中心时安装木方及螺旋顶杠就位，拧紧顶杠，该打木楔处打上木楔 Bottom lining up to slightly higher than kiln center, put beams and screw jacks against brickwork, tighten the screw jack, hammer in wooden wedges between the brick hot face and the wooden beam
	继续耐火砖铺底，最终一环一环地把耐火砖锁紧 Continue bottom lining, finally close brick rings one by one

在窑筒体下部衬砌作业，锁砖也在下部，需要慢转窑两次。

Lining is always done at the bottom of the shell, and keying bricks are also placed at the bottom of the shell. The kiln needs to slowly rotate 2 times.

螺栓（槽钢）方法
Bolt (channel) method

螺栓槽钢法需要熟练的电焊工

Bolt (channel) method needs skillful welders

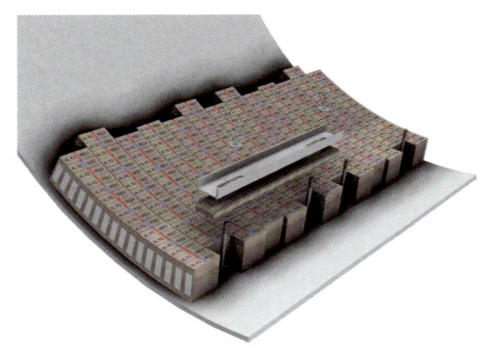

特制的槽钢和木方

Special made channel and wooden beam

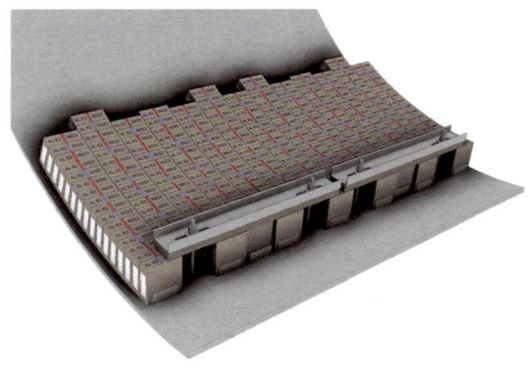

在朝下的砖侧面焊小钢板，有时还需要在木方和砖热面之间打入小木楔以确保安全。

Weld steel sheets on brick side faces towards the downward direction, and sometimes small wooden wedges need to be hammered into the joints between the brick hot face and the wooden beam to ensure safety.

交错砌筑时采用螺栓法砌筑
Use the bolt method for compound lining

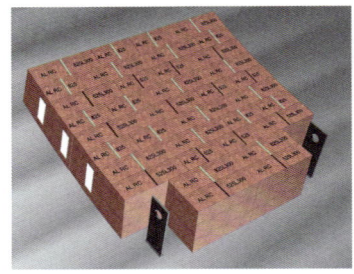
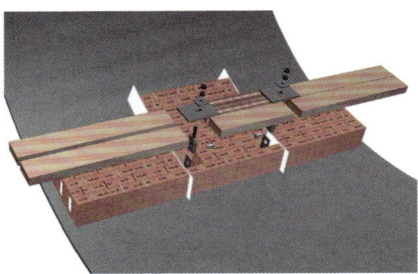

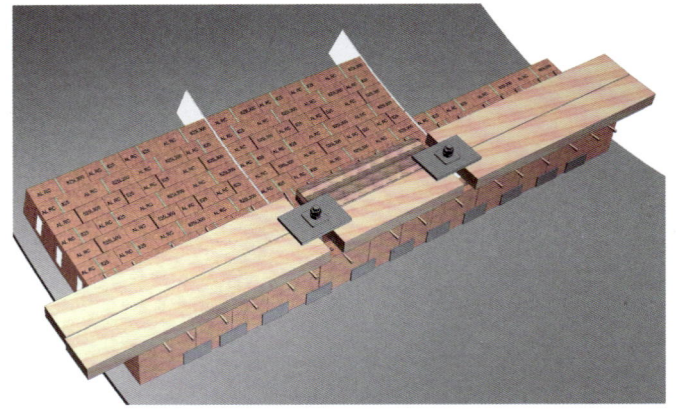

对于窑筒体变形区域或者窑口区域,有时需要采用交错砌筑,螺栓法衬砌是可操作性强且简便的方法。

For shell deformed area or outlet area, sometimes need to use compound lining patten, bolt method would be a more applicable and simple lining way.

其他方法
Other methods

随着回转窑直径的不断增大,砌砖机法已经成为主流砌筑方法,这是目前最安全且最高效的砌筑方法,随后内容会详细述及。胶结法需要对窑筒体喷砂除锈并保持窑筒体干燥,且存在粘胶过期问题。弹簧撑杆法需要很多弹簧撑杆及一木拱架。这两种方法在此不再介绍。

As the kiln diameter increases, the lining machine has become the mainstream and safest method for lining cement rotary kilns due to its efficiency. More details will be presented later. The glue method requires sandblasting to remove rust from the shell and keep it dry. Expired glue cannot be used and will not be discussed here. The pogo stick method involves many pogo sticks and a wooden arch rig, but these two methods will not be discussed further in this context.

砖型的选择
Brick shape selection

衬砌厚度 (mm) Lining height(mm)	回转窑直径 (m) Diameter of rotary kiln (m)
180	< 3.6
200	> 3.6~4.3
220	> 4.3~5.4
250	> 5.4

* 一般性建议,可根据实际情况予以调整。

* This is a general recommendation and can be adjusted according to actual requirements.

普遍性的建议：回转窑内碱性砖使用 VDZ 砖型砌筑（可干砌亦可湿砌），非碱性砖使用 ISO 砖型湿砌。

General recommendation: VDZ is applied basic brick lining and ISO shapes for alumina bricks.

使用 VDZ 砖型衬砌，砖缝约多了 30%, 这样砖环的柔性更好。

VDZ linings have approximately 30% more brick joints, providing increased flexibility and thermal shock resistance to the rings.

对于直径很大的窑（$\phi>5.6m$），建议碱性砖使用 ISO 砖型，因为 ISO 砖型的锥角更大，砖衬稳定性会更好。

For kilns with a very large diameter ($\phi>5.6m$), ISO shapes are also recommended for basic bricks. This is because of the greater taper, which increases lining stability during kiln start-up.

大小头尺寸之差
Cold face and hot face

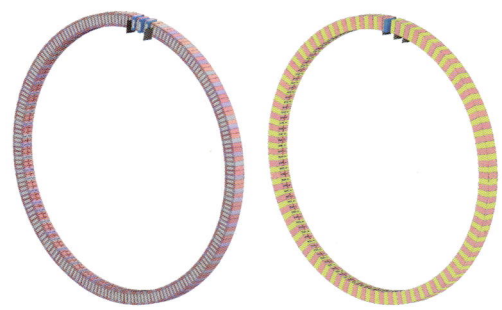

  VDZ（德国水泥协会）和 ISO（国际标准组织）砖型需要两种标准砖型以特定的配比来衬砌某一直径的水泥回转窑，一般还需要 2 型锁缝砖来封闭砖环。注意 VDZ 和 ISO 标准中都没有规定锁缝砖砖型，锁缝砖砖型是由供应商规定的。衬砌配比是基于窑筒体是正圆的假设计算出来的，在砌筑过程中，衬砌配比应根据窑筒体的形状做出调整。

  Both VDZ (Verein Deutscher Zementwerke, the German Cement Works Association) and ISO (International Organization for Standardization) shapes require 2 standard shapes mixed in a certain ratio to line a cement rotary kiln of a specific diameter. Typically, 2 key bricks of various shapes are needed to complete the brick ring closure. It's important to note that the key shapes are not specified in either the VDZ or ISO standards but are determined by brick suppliers. The mixing ratio is calculated based on the assumption that the kiln shell is round. During lining, the mixing ratio should be adjusted according to the actual shape of the shell.

火泥
Mortar

微不足道成就不凡
Minor creates great

古罗马引水渠
Roman Aqueducts

在砂浆发明之前,石质建筑物的石料被加工成规则的矩形,而后一块一块摆上去。

Before mortar was invented, buildings were made from rectangular stones that were chiseled to be flush, and then they were placed one by one.

图示水泥回转窑耐火砖砌筑

古罗马人用火山灰和石灰发明了罗马水泥，发明了砂浆和罗马混凝土。

The ancient Romans invented Roman cement using pozzolan and lime, which led to the invention of mortar and Roman concrete.

使用砂浆和混凝土，不规则形状的石料就可以利用了。微不足道、其貌不扬的材料成就了古罗马建筑的不凡。

With mortar and concrete, irregularly shaped stones could be used, and this minor substance helped create many immortal Roman historic relics.

水泥回转窑的砌筑是依照特定规范施工的瓦工工作，需要传统瓦工工器具和材料，传统的瓦工工作方法和技能依然可以在窑的砌筑工作中实施，不过必须按照回转窑砌筑的特定规范来进行。

Lining a cement rotary kiln involves masonry work based on specified installation regulations. It requires conventional masonry tools and materials, and traditional masonry methods and skills can still be used but must adhere to the specified regulations.

火泥在耐火砖砌筑过程中依然不可或缺，砌筑中用于窑筒体变形处的调整调节，补偿火砖外形公差，湿砌增加砖环柔韧性，封闭砖缝防止窜气等。

Mortar application is essential throughout the entire brick lining process of a cement rotary kiln. It helps in adjusting bricks in deformed shell areas, compensating for brick tolerance, improving the flexibility of the brick ring, closing brick joints to prevent hot gas penetration, and so on.

干砌依然需要用火泥调整
Clench lining still need mortar adjusting

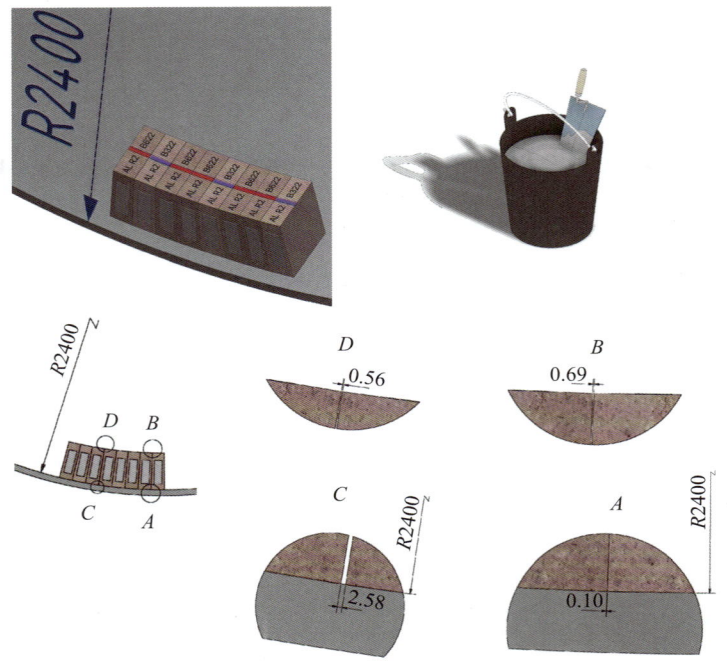

即使完全按照正圆计算出来的砖型配比，砖型按照配比组合起来时还是会出现大的砖缝，干砌时依然需要用火泥进行调整。

Even if the calculated theoretical mixing ratio, which is based on the shell being a round circle, results in larger brick joints, mortar is still needed to adjust or compensate.

还经常需要调整砖型配比来保证"坐牢，贴紧"。

Often, it's necessary to adjust the mixing ratio to ensure a "firm seating on the shell and a tight fit of each side face".

通过火泥可以保证耐火砖"坐牢在窑筒体上，大面贴紧"。

Mortar is the material that ensures the bricks "sit firmly on the shell and have a tight fit on each side face".

在窑变径处只能湿砌
Cone section only can be lined with mortar

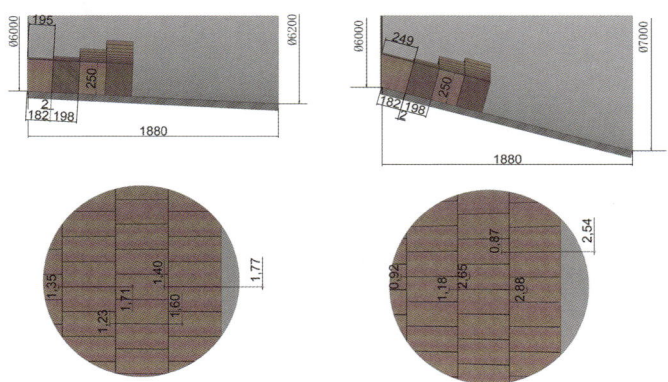

轮带区域，建议湿砌
Tyre section, mortar lining is recommended

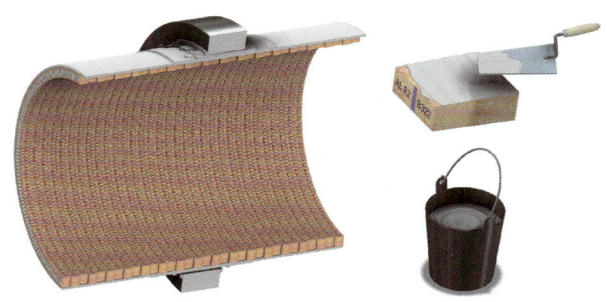

非碱性砖区域，只能湿砌

None basic brick zone, only mortar lining

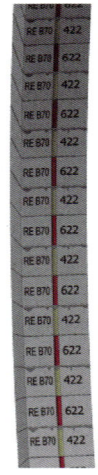

不推荐但可以接受在砌筑过程中使用薄钢片来进行调整

Not recommend, but it's acceptable to use thin steel shims to adjust bricks during lining

干砌可以高效且干净，不过湿砌是完美的砌筑方法。一般碱性砖采用干砌，非碱性砖采用湿砌。

Clench lining could be more efficient and clean, but perfect lining always be done by mortar lining. Normally basic brick lined in clench way and non basic brick lined with mortar lining.

（耐火砖上贴的）纸板和钢板
Cardboards & Metal sheets (attached on brick)

纸板（碱性耐火砖）

环向砖缝处贴有纸板（1.5~2mm 厚），用于补偿轴向热膨胀。注意在投料之前，点火升温不要中断。

钢板（MA，镁铬砖）

钢板（0.7mm 厚）贴在轴向砖缝处，600°C 许软化，当温度超过 1000°C 时氧化，在砖之间形成陶瓷结合，当锁砖打入钢板时，贴着的钢板需要剥下。

如今，这种贴钢板的耐火砖已几乎过时。

Cardboards (CB, basic bricks)

Attached to the radial joint (thickness 1.5-2 mm), for compensating brickwork axial thermal expansion, heating up cannot be interrupted before feeding in.

Metal sheets (MA, magnesia chromite bricks)

Attached to the axial joint of each brick (thickness 0.7 mm). It starts to soften at approximately 600°C and oxidizes at temperatures exceeding 1000 °C, forming a ceramic bond between bricks. When inserting the key shim, remove the attached metal shim.

Nowadays, very few demand for such bricks.

在投料之前升温中断了,纸板被烧掉,再次升温后,再次开窑和投料后,在中间挡砖券之前,形成了一条不规则的缝隙。

Heating-up interrupted before feeding, cardboard had been burnt off, when kiln restarts and feeding, an irregular gap in front of middle brick retaining ring formed.

在窑筒体上画出轴向（基准）线和环向（基准）线
Marking the axial and radial reference lines

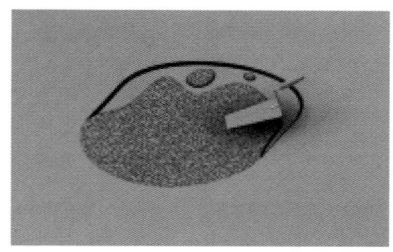

画基准线前清理好窑筒体。窑门处于正下方，用浇注料找平
Clean the shell before marking the reference line. The kiln manhole is located below, and it should be paved flush with the shell using castable

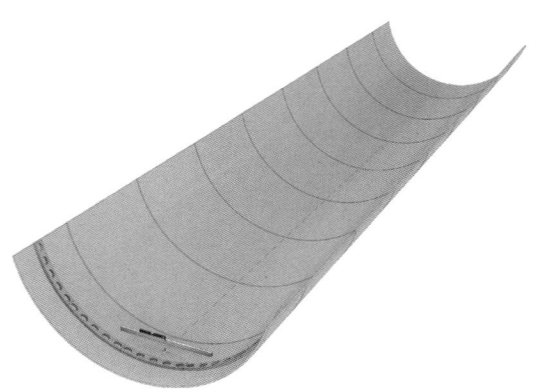

使用水平尺和铅坠画出轴向基准线
Use a level and plumb bob to fix the axial line

画环向线需要一个长圆规

To draw an accurate radial reference lines needs a long compass

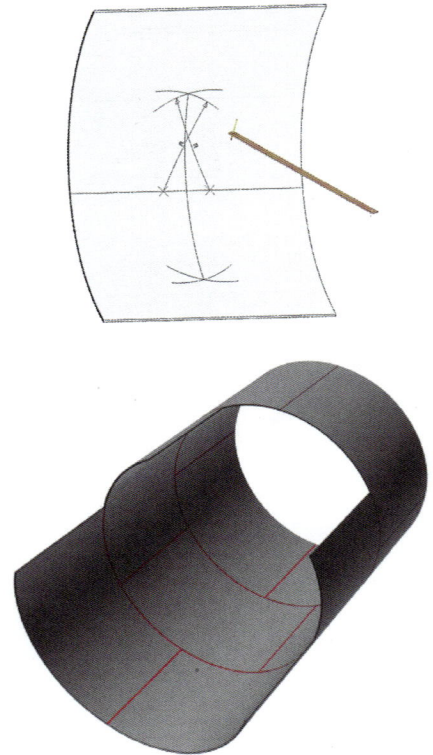

使用圆规画出环向基准线或者就是以窑筒体环向焊缝来做环向基准线。

Use long compasses to draw radial reference lines, or you can simply use the radial welding seam as your radial reference lines.

窑筒体的段节端面的焊接坡口是在车床上加工出来的，在砌筑过程中，环向焊缝可以粗略地作为环向参考线。

The end of the kiln shell section is machined by a lathe to create a welding groove. so the radial welding seam could be serve as a rough radial reference line during lining.

砖的排布模式
Lining patterns

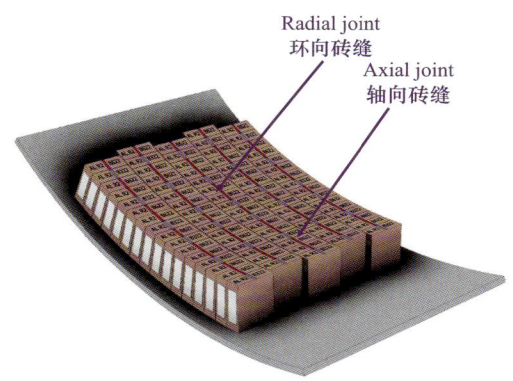

推荐：错缝砌法（花砌）
Recommended: Staggered lining pattern

不推荐

Not recommended

交错砌法

Compound lining pattern

直缝砌法

Straight lining pattern

砌筑开始
Start lining

大型窑一般和窑口挡砖圈接触的几环砖采用高强度的非碱性砖，用以抵抗砖衬推力。

For large-diameter kilns, it's common to line several rings of high-strength non-basic bricks against the outlet brick retaining ring to resist lining thrust.

小型窑窑口挡砖圈处直接衬砌碱性砖，打浇注料之前，砖需要用塑料布覆盖或者沥青涂敷，防止碱性砖接触到浇注料中的水分。

For small-diameter kilns, basic bricks are lined directly against the outlet brick retaining ring. Before installing castable, the 1st basic brick ring should be covered with plastic film or coated with bitumen to prevent moisture from fresh castable.

从轴向基准线开始砌砖
Start lining from the axial reference line

耐火砖铺底
Bottom lining

耐火砖摆放好后用皮（木）槌敲打。

要不时地检查砖环的环向面是否平齐。

不时地检查火砖是否："坐牢"在窑筒体上，火砖各面相互"贴紧"。

When placing bricks on the shell, tap them gently with a rubber (wood) mallet (hamer).

Frequently check if the radial surface of the ring is flush.

From time to time, check whether cold face of brick(two edges) "sit on kiln shell firmly", the side faces of brick "contact each other".

窑筒体上的焊缝
Welding seams of kiln shell

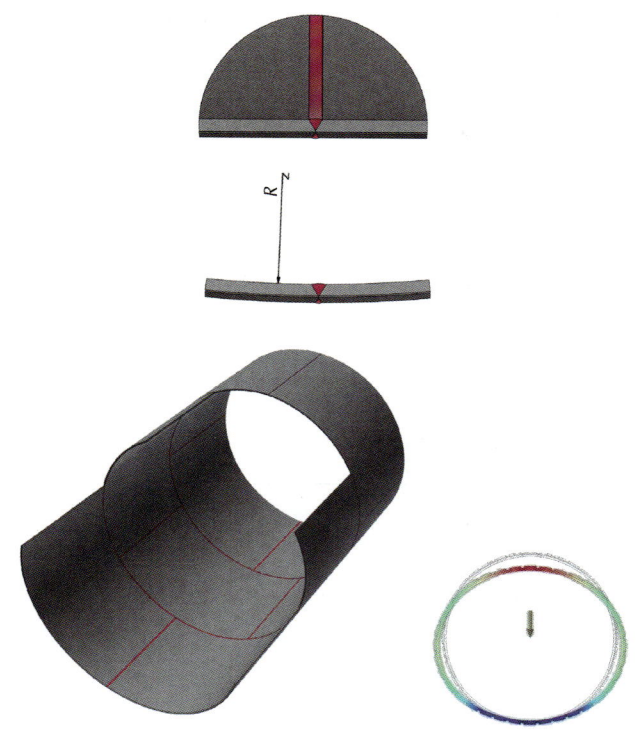

注意窑筒体上既有轴向焊缝,又有环向焊缝。焊缝往往会高出窑筒体,焊缝处的窑筒体还会因为焊接产生变形,砌筑时砖型配比往往在焊缝处需要调整,而且经常需要用火泥进行适当调整。

另外,窑筒体还会因自重产生一定的变形,所以砌筑时,要根据窑筒体的情况调整砖型配比。

Note that the kiln shell has both axial and radial seams. Typically, welding seams are not flush with the kiln shell; they are slightly higher than the shell surface. Additionally, adjacent to welding seams, the shell may have some deformation due to welding, and the mixing ratio of the welding seam nearby often needs to be adjusted accordingly. This adjustment is done through mortar lining.

Furthermore, the shell could be deformed by its own weight, so the mixing ratio should be adjusted based on the shape of the shell.

骑在环向焊缝处的砖环
Brick ring on welding seam

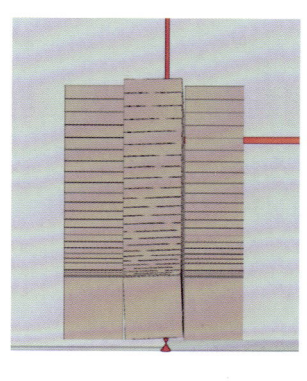

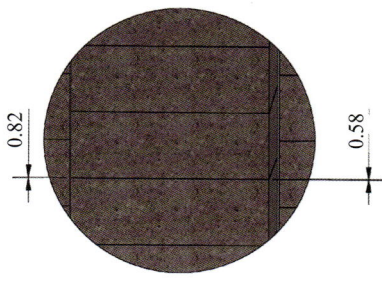

骑在环向焊缝处的砖环呈现出直径大的一端砖缝大（直径大些），另一端砖缝小（直径小些）。每放置几块就需要用火泥来调整，不然砖环会歪斜。

The brick ring on the weld seam shows that one side has bigger joints (a larger diameter), while the other side has smaller joints (a smaller diameter). So, after placing several bricks, mortar needs to be applied to make adjustments, otherwise, the brick ring will be skewed.

耐火砖面打火泥
Mortar applied on bricks

碱性耐火砖干砌调整时砖面打火泥
Mortar applied for adjusting during basic brick clench lining

非碱性耐火砖湿砌时砖面打火泥
Mortar applied for non-basic brick mortar lining

一般火泥厚度为 1~2mm。

Normally, the thickness of mortar applied on brick surface is between 1-2mm.

砖环歪斜——侧视
Skewed brick ring—side view

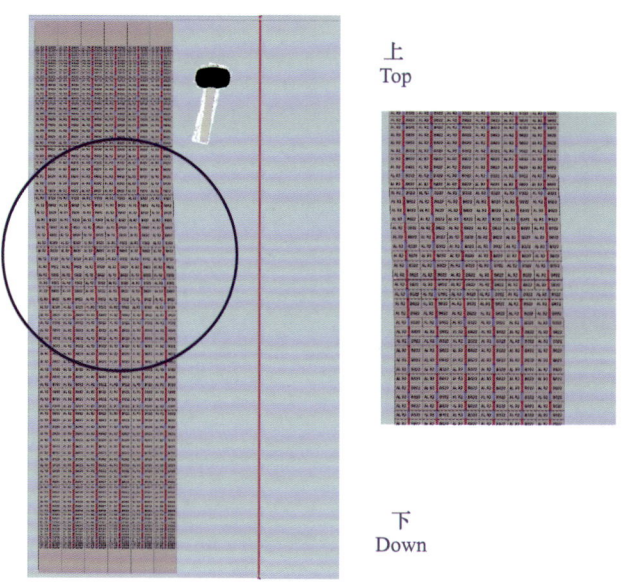

上 Top

下 Down

铺底时下部砖的径向端面敲打力度较小，上部砖的径向端面敲打力度较大。

Less or no rubber hammer knocking is used for the bottom lining, while heavy rubber hammer knocking is used for the upper lining.

砖环歪斜——下视
Skewed brick ring—Top view

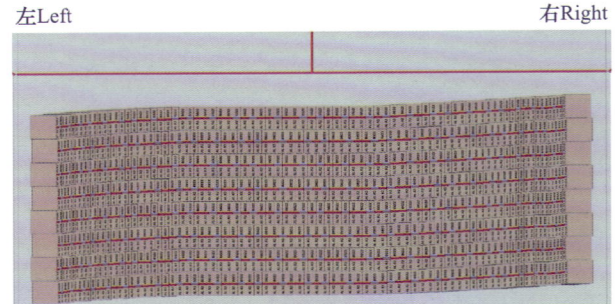

两侧砌筑时砖的径向端面敲打力度不一致,还可能一侧敲打而另一侧没有敲打。

Different rubber hammer knocking on both sides, even one side knocked and another side no knocking.

如何避免砖环歪斜
How to avoid skewed ring

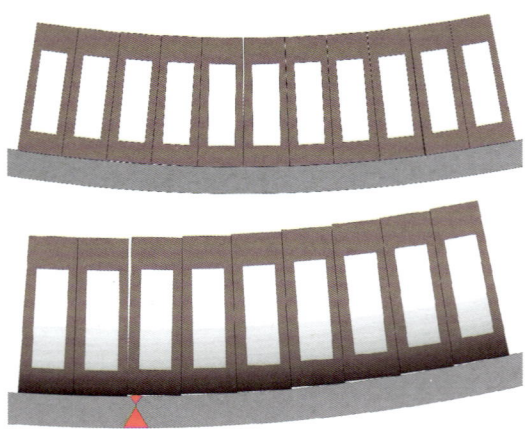

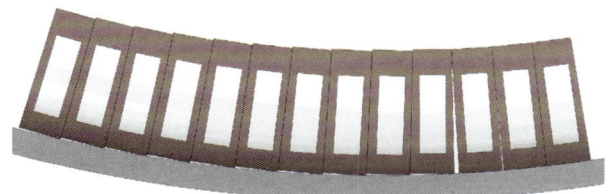

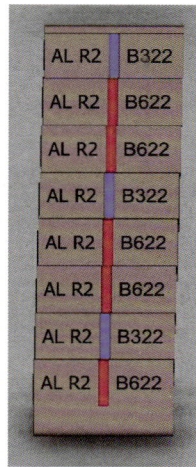
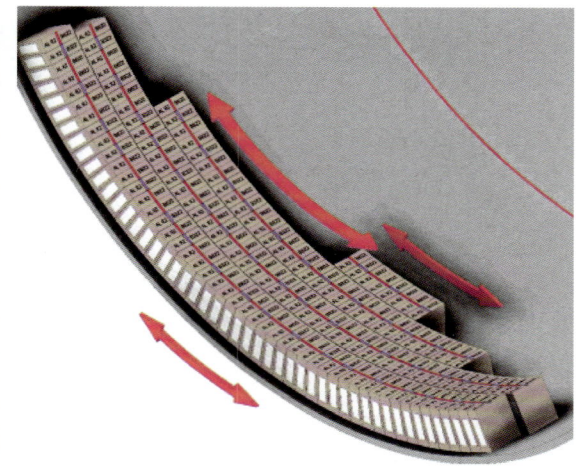

砖环径向端面和冷面表面（箭头方向）不得有连续的台阶（不能连续一个方向出台）。

砌筑每接近窑筒体的环向焊缝时，检查砖环两侧上下左右和焊缝的距离。

Ensure that the radial face and cold faces of the brick ring are flush (indicated by the arrow direction). There should be no continuous steps. When the brick ring is close to the radial welding seam, check the distance between the brick ring and the seam.

耐火砖面打火泥

砖环的锁紧程度

Brick ring tightness

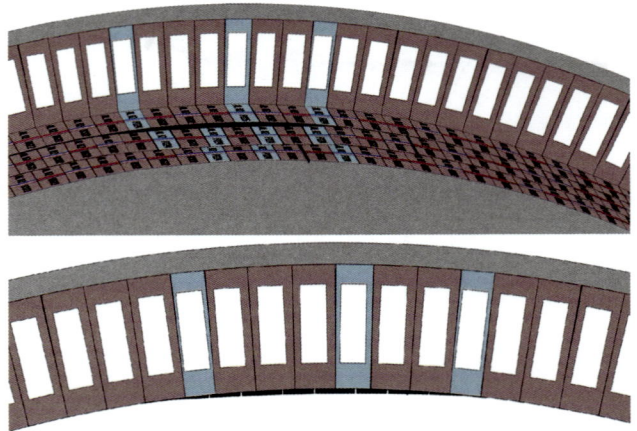

砌砖机的撑脚（或螺旋顶杠）松开后，砖环应该依然紧贴窑筒体，如果砖环向下沉（如上图），则应补加钢板，此为砌筑检查要点。非碱性砖砖环要比碱性砖砖环更紧些。

After releasing the bricking rig (or screw jacks), the rings must remain in contact with the kiln shell. If the ring sags (as illustrated above), add additional closure shims. Non-basic brick rings should be tighter than basic brick rings.

锁缝砖和钢板的布置
Arrangement of key bricks and closure shims

规范
Regular

尺寸小于标准砖的锁缝砖（蓝色）用标准砖间隔开。钢板打在标准砖侧面。如果锁缝砖尺寸大于标准砖，则不受此约束。

一块砖不允许两面都打入钢板。

Key bricks with dimensions less than standard bricks should be divided by standard bricks. Closure shims should be driven into the side (axial joint) of standard bricks. If the key brick is larger than the standard brick, there is no above mentioned limit.

It is not allowed to drive closure shims into both sides of one brick.

不规范
Irregular

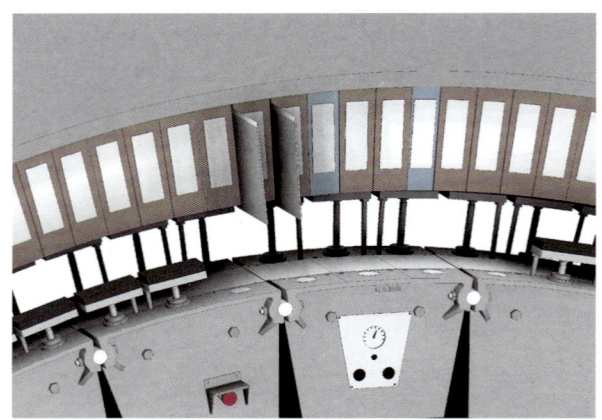

错误
Wrong

* 不规范指的是发生了，不必拆除，但不允许再次发生。错误则必须拆除返工。

* Unregular means if happens, no need remove, but not allowed happens again. Wrong means must be removed and reinstall.

砖环封闭时的要点
Key points when close brick ring

避免（径向或者轴向）楔形缝隙出现

Avoid (radial or axial) wedge joints

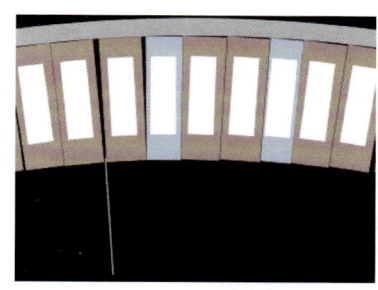

不规范
Irregular

不规范
Irregular

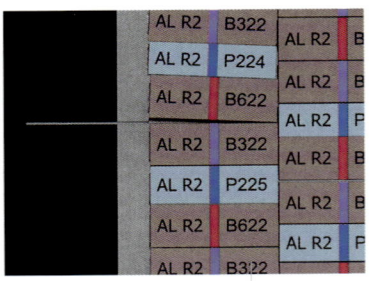

不规范
Irregular

封闭砖环之前，在砖的大面打火泥，可以避免楔形缝隙的出现。

Before closing brick ring, apply mortar on brick side face, can avoid such wedge joints.

不能用铁锤直接打击耐火砖
NEVER use metal hammer to drive in brick

正确
Correct

错误
Wrong

把最后一块火砖插入：如果尺寸恰当，没有敞开砖缝，可用铁锤垫木方打入，铁锤决不能直接打击火砖。

When inserting the last brick and the brick combination is perfect without any open joints, AVOID knocking with a metal hammer onto the brick DIRECTLY.

打进锁砖钢板
Drive in closure shims

建议采用气铲（锤）打入锁缝钢板，注意气压必须高于 6.5bar（1bar=0.1MPa）。气动锤可以打入 2mm 的钢板。

Recommend to drive in closure shim with pneumatic hammer, notice pressure should reach 6.5bar (1bar=0.1MPa). 2mm closure shim can be driven by pneumatic hammer.

也可以使用铁锤打入钢板，一般情况下，铁锤更容易将 2mm 的锁缝钢板打弯，更多情况下是使用 3mm 锁缝钢板时使用铁锤打入。

Metal hammer also can be used to drive in closure shim, normally, metal hammer tends to knock the 2mm closure shim bending, in most case, metal hammer for driving in 3mm closure shim.

锁缝钢板和自制锁缝钢板夹持器

Closure shim and self-made closure shim holder

两边开刃的锁缝钢板，边部尺寸小于砖的边长 10mm。材料使用中国 Q235 碳素钢钢号。

The closure shim has two sharp edges and its side dimensions are 10mm smaller than the brick side dimensions. The material used is Q235 grade carbon steel (China).

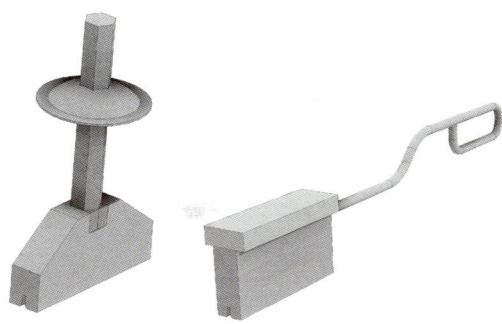

自制钢板加持具有助于提高效率和安全
The self-made closure shim holder can help improve efficiency and safety

锁缝砖和锁缝钢板纵向布置

Arrangement of key bricks and closure shims longitudinally

锁缝砖和锁缝钢板不要放置在一条直线上

Position the key bricks and closure shims in a staggered manner

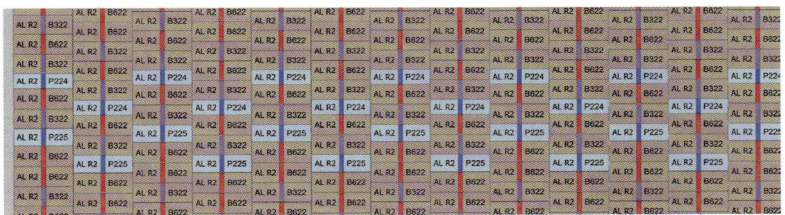

规范
Regulated

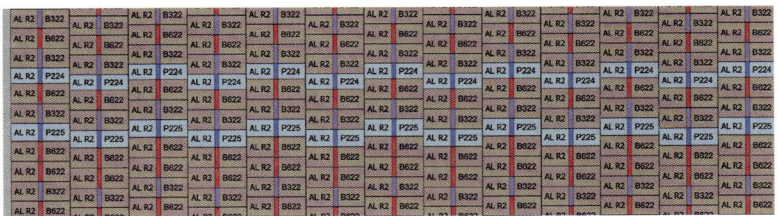

不规范
Irregular

新旧砖交接
The new lining meets the old lining

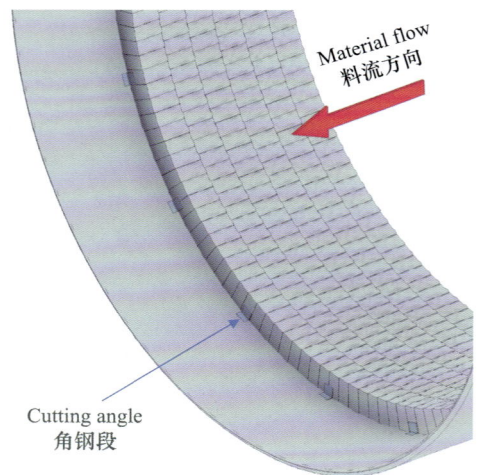

Cutting angle
角钢段

Material flow
料流方向

　　拆除旧火砖后,保留的旧火砖要用焊接在窑筒体上的角钢固定住。角钢段节紧贴着旧砖砖环焊接,防止旧砖砖构在慢转窑时向下窜动。

After removing the worn bricks, the remaining old lining in the kiln needs to be fixed with angles welded onto the kiln shell. Weld the cutting angle against the remaining used brick ring to prevent the brickwork from moving downwards when the kiln rotates slowly.

　　旧砖环径向砖面上经常会粘结有块状硬料,应当打磨或者刮掉,不要用铁锤敲打。而且旧砖环径向砖面经常是不平整的,需要用火泥找平。

Often some hard material stick on radial surface of remained used brick ring, should be ground off or scraped off, not use hammer to knock out. Furthermore, the radial surface of remained used brick ring is often not flush, need to be applied flush with mortar.

新旧砖交接时耐火砖的加工

Brick cutting when new lining meets old lining

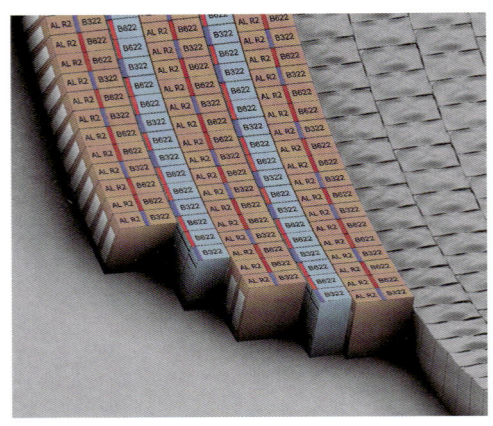

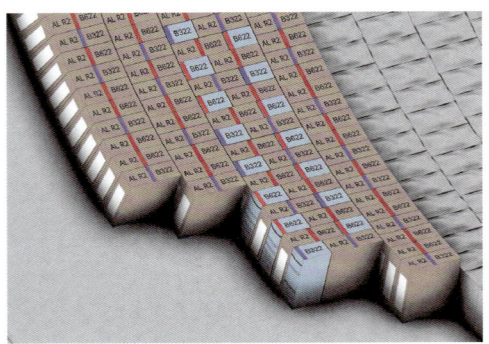

和旧砖接触的一环用整砖。

如果需要加工两环砖，两环加工砖用整砖间隔开。

如果新旧砖间的间隙为200~300mm，最好使用长砖。如果需要切标准砖，切砖尺寸不要小于100mm。

还可以使用标准砖和加工砖交错方法来增强砖环的稳定性，可以人为地把两环变成一环。

A full complete brick ring contact with old brick ring, install cut bricks between full rings.

If need to cut 2 rings, separate the two cutting rings with a full brick ring.

Use long bricks (L shapes) to fill gap of 200-300mm, never use bricks <100mm.

Compound lining enhances stability by artificially combining two rings into one.

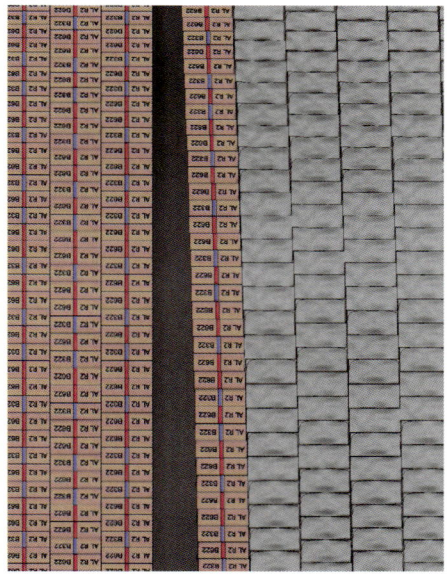

经常遇到这种费时费力的情况，加工砖逐块去测量，逐块加工。

We often encounter such time and energy consuming cases where each cutting brick needs to be measured and cut individually.

不允许带水切割镁质砖

NEVER to cut magnesia bricks with water

局部挖补耐火砖

Partial repairs (patching)

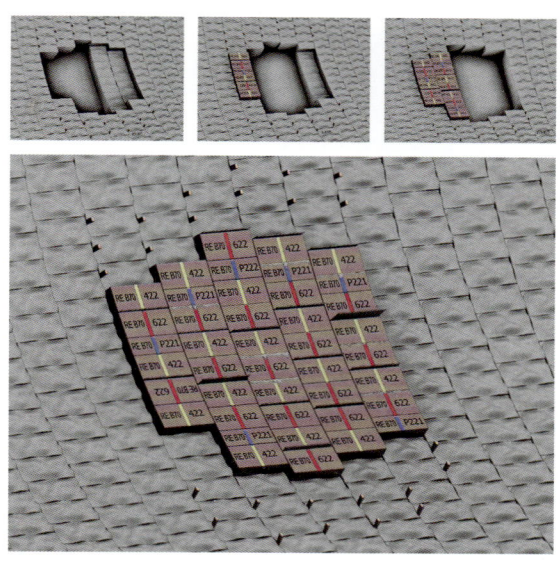

注意安全！

拆砖之前用木楔固定好要拆的砖环；

拆砖时不要拆过三环；

挖补总是需要更多的火泥和锁砖钢板。

Caution!

Before removing the replaced bricks, use wood wedges to fix the brick rings to avoid brick rings slip down；

Open the brick rings no more than 3 rings；

Patch repairing ALWAYS needs more mortar and closure shims.

窑筒体变形，无法干砌

Shell deformed, clench lining cannot proceed

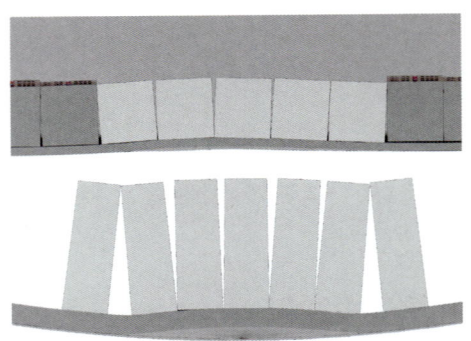

窑筒体变形——湿砌

Shell deformation—mortar lining

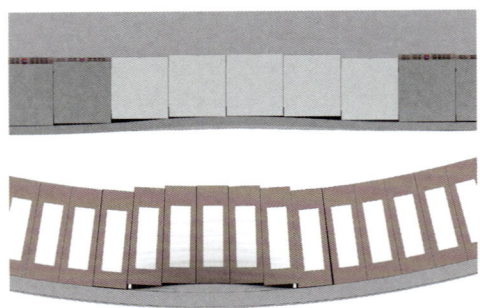

在变形过大之处可以垫小钢片支承并填充火泥。

Place steel shims to support the brick in the most deformed position, and fill the gap with mortar.

如果变形超过 8mm，则要考虑修补或者更换窑筒体了。

If shell deformation exceeds 8mm, please consider repairing or replacing the shell.

在没有对正的新换的窑筒体上砌筑
Lining on misaligned replaced shell

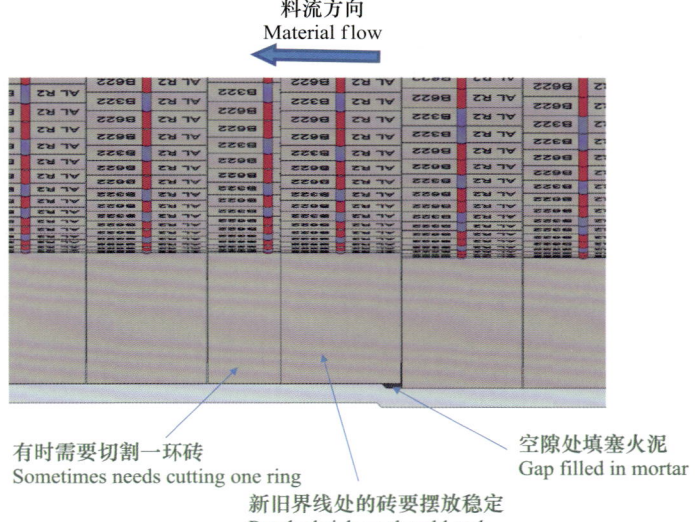

料流方向
Material flow

有时需要切割一环砖
Sometimes needs cutting one ring

新旧界线处的砖要摆放稳定
Put the brick on the old and new shell boundary stable

空隙处填塞火泥
Gap filled in mortar

挡砖圈，窑口

Brick retaining ring, outlet

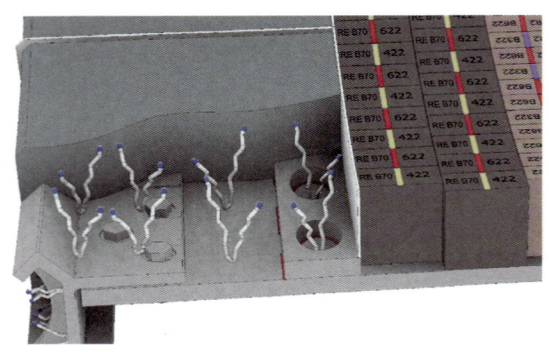

关于挡砖圈的设置有不同的观点：

欧洲：除窑口外，窑内不设挡砖圈，取决于设计，挡砖券沿圆周方向每 30°一段。

日本：窑内设多道挡砖圈。

常见做法：

窑口一道挡砖圈，碱性区域不设挡砖圈，非碱性区域设置一道挡砖圈。

无论如何，每段挡砖圈的角度不大于 30°，一定要钳工找正，焊工焊接，要垂直于窑轴线，平齐，和耐火砖接触区域无焊缝凸起。

There are different viewpoints regarding brick retaining rings:

Europe: only outlet brick retaining ring, no more in kiln, normally each 30° one section.

Japan: several brick retaining rings arranged inside kiln.

Common application:

The outlet has one brick retaining ring, and there is no brick retaining ring in the basic zone. However, the non-basic zone has one brick retaining ring.

In any case, each section of the brick retaining ring should be ⩽ 30°, aligned and welded professionally, perpendicular to the kiln axis, and all sections should be flush.

No protruded welding seam on the side contact with brick ring.

Anyway, each brick retaining ring section ⩽ 30°, should be aligned and welded by professional, perpendicular to kiln axis, and all sections are flush.

挡砖圈——窑内
Brick retaining ring—middle

用浇注料和陶瓷纤维棉填塞，避免出现任何空隙
Fill in castable and ceramic fiber, avoid any gap

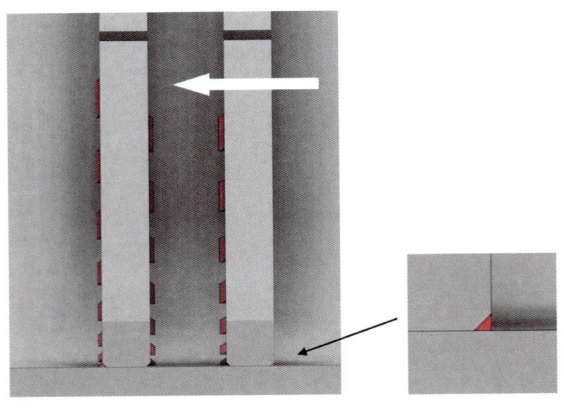

窑筒体会被窜入双道挡砖圈环形缝隙内的颗粒状物料磨穿。

Shell could be worn through by material particles that penetrated inside the ring channel of double brick retaining ring.

单道挡砖圈的形式
Single brick retaining ring-form

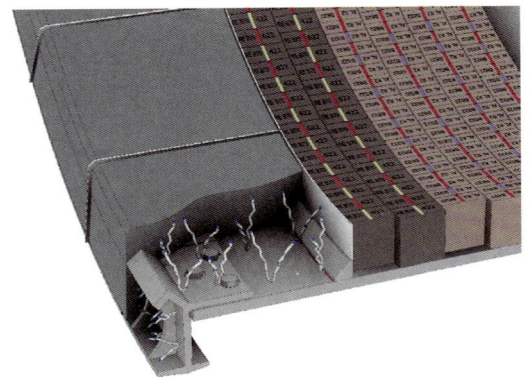

注意中档挡砖圈和下游砖环之间留有一定的热膨胀补偿间隙，该间隙要用陶瓷纤维棉填塞住。

Please note that between the middle brick retaining ring and the downstream adjacent brick ring, you should leave a gap for heat expansion. This gap must be filled with ceramic fiber.

窑变径处砌筑
Brick lining on the cone

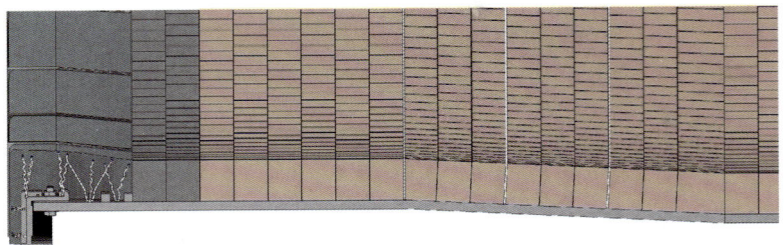

大型回转窑为了减缓窑口处的砖衬推力，经常会设置窑口变径。

放置在窑变径砖环之间的陶瓷纤维毡不仅仅起到膨胀补偿的作用，还起到砖构因砖衬推力移动补偿的作用，衬砌在窑变径处的耐火砖必须湿砌。

衬砌在窑变径处的耐火砖既可以使用特殊设计的异型砖（不需要再加工），也可以使用标准砖，使用标准砖时，第一环和最后一环需要切割加工。

For resisting high lining thrust, large kilns often arrange one cone in outlet.

The felt placed between the brick rings on the cone serves not only for thermal expansion compensation but also for compensating for brick displacement (as per the design) due to lining thrust. The brickwork on the cone must be mortar lining.

The brick on the cone can use special shape, no need cutting, or standard shape brick with 1st and last ring need to be cutting.

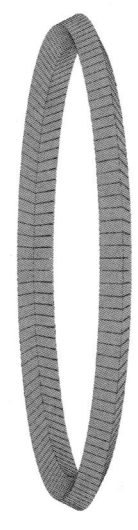
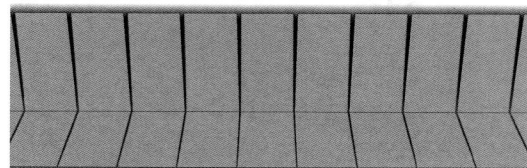

注意圆锥区域的耐火砖砖环径向端面是个多棱锥，近似于圆锥面，轴向砖缝呈楔形。故变径圆锥位置必须湿砌。

Note that the radial surface of the brick ring on the cone is not a flush plane; it is actually a pyramid, resembling a cone surface. The axial brick joints form a wedge shape. Therefore, mortar lining on the cone is necessary.

使用 DAT 砌砖机进行砌筑
Lining with DAT lining machine

最安全高效的砌筑方法

The most safe, efficient lining way

正常情况下，对于 ϕ4.8m 回转窑，砌筑速度为 0.6~1m/h（3~5 环 /h）。

Typically, for a kiln with a diameter of 4.8 meters, the lining speed varies from 0.6 to 1 meter per hour (3 to 5 rings per hour).

拱架的进化 → DAT 砌砖机

Evolution of arch method → DAT lining machine

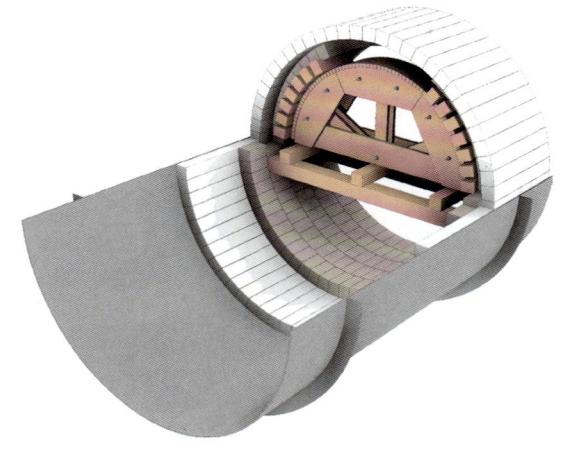

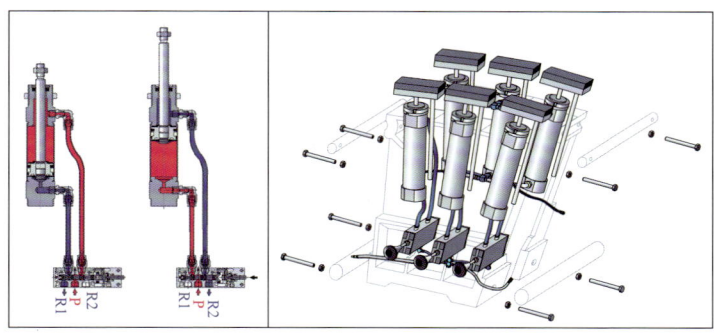

木楔变成了气缸。

Wooden wedges have been replaced by pneumatic cylinders.

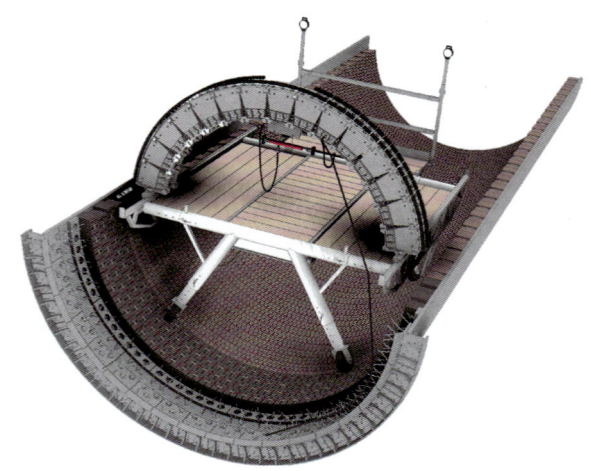

木拱架变成了铝合金模块组合拱架。

The wood arch has been replaced by an aluminum alloy pneumatic module arch.

遵循砌砖机作业指导来装配砌砖机！

FOLLOW the instructions in the DAT operating manual to assemble it!

按照砌砖机操作手册，把固定销插入砌砖机腿和臂上正确的孔中。

确保固定销牢固固定！不然，会造成事故！

Follow the DAT operating manual instruction, push the fixing pins in correct holes on telescopic legs and arms.

MAKE SURE the fixing pins secured! OR, it could result in an accident!

DAT 砌砖机的装配——第一步
Assemble the DAT lining machine— Step 1

人工竖起装配好的前后门架。

Erect the assembled front and rear portal frame manually.

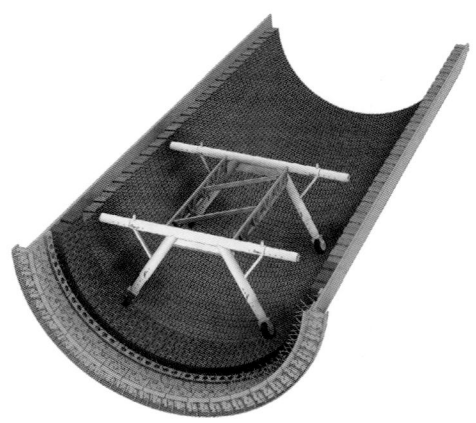

装上中间支撑构件和斜撑。

Install middle bracing and diagonal bracings.

DAT 砌砖机的装配——第二步

Assemble the DAT lining machine— Step 2

装好两侧的导轨支架，拧紧螺钉。

Position the rail supports on both sides and tighten the fixing screws.

装好中间支撑杆和外侧支撑杆。

Middle bracings and outside bracings.

DAT 砌砖机的装配——第三步
Assemble the DAT lining machine－Step 3

摆放木板,形成稳定的工作平台。

Place the planks to form the stable working platform.

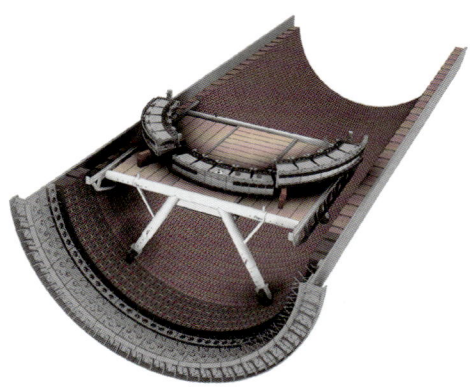

在砌砖机上的工作平台组装砌砖机模块拱架,连通拱架内气路。

Assemble the segmented arch on the working platform and connect the pneumatic circuit pipes within the arch.

DAT 砌砖机的装配——第四步
Assemble the DAT lining machine— Step 4

把砌砖机小车轮子放在轨道上，用人力或者绞车竖起整个砌砖机模块拱架。

Place the lining machine barrow wheels on rail, erect the assembled full DAT segments arch manually or by hoist.

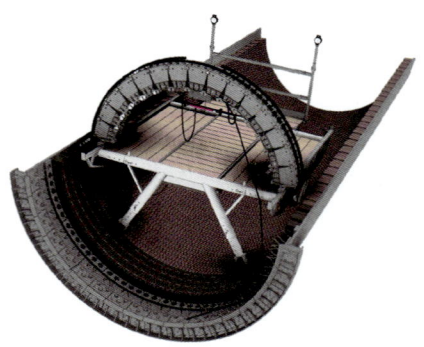

把砌筑工具放置到位，接通外部压缩空气，打开照明。

Place all lining tools on the machine, connect the compressed air supply, and switch on the illumination.

DAT 砌砖机砌筑

Lining with DAT machine

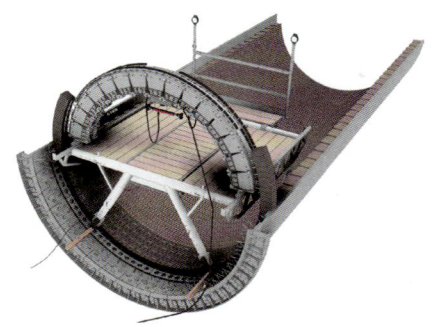

第一环向上砌筑开始
Start the first ring upwards

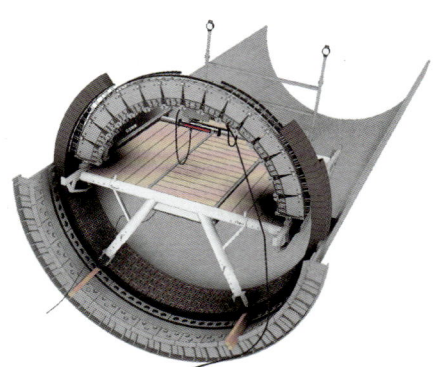

第二环向上砌筑开始
Start the second ring upwards

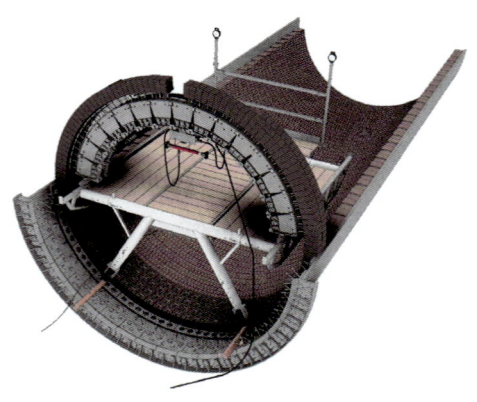

第一环继续
Continue with the first ring

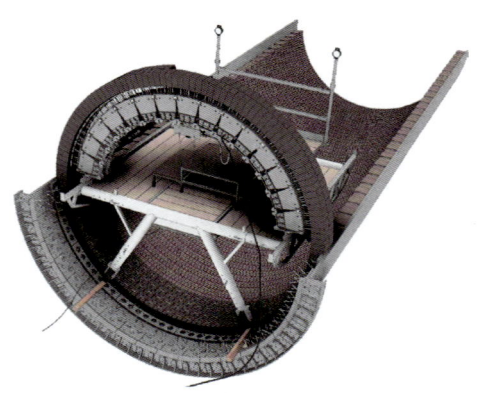

第一环锁口封闭，第二环即将锁口
The first ring is closed, and the second ring is ready for closing

用液压千斤顶顶紧第二环砖，放下气缸橡皮撑脚，移动气动拱架一环砖的距离。

Tighten 2nd ring opening by hydraulic jack, release all supporting rubber feet of pneumatic cylinders，move the arch one ring distance.

使用液压千斤顶自砖环开口处顶紧第二环砖
Use a hydraulic jack to tighten the second ring in the ring opening

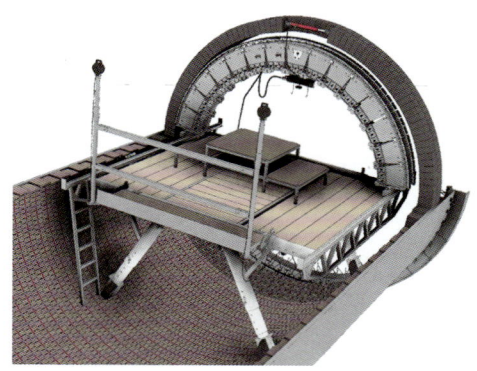

放下所有的气缸橡皮撑脚，移动气动拱架一环砖的距离
Withdraw all pneumatic cylinder rubber feet, moving the pneumatic arch one ring distance

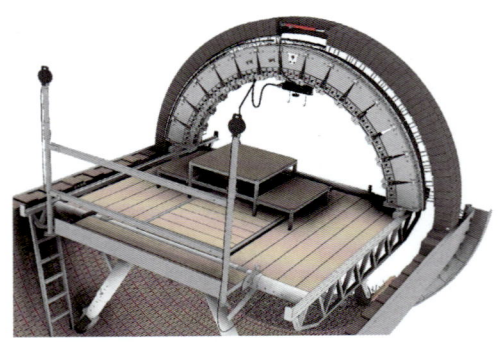

用第一排气缸橡皮撑脚顶住第二环砖
Press the first row of rubber feet from the pneumatic cylinders against the bricks of the second ring

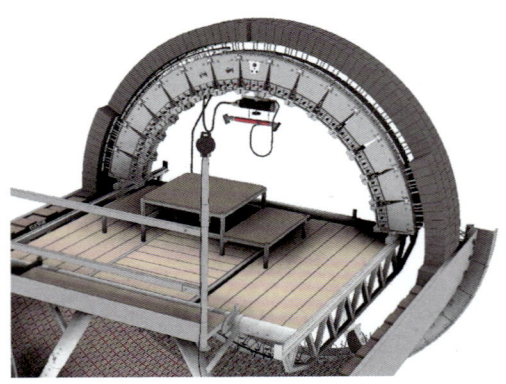

第二环砖锁口
Close second ring

重复上述操作

Repeat the above operation

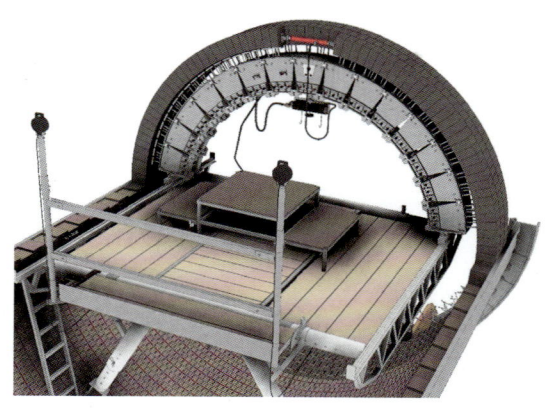

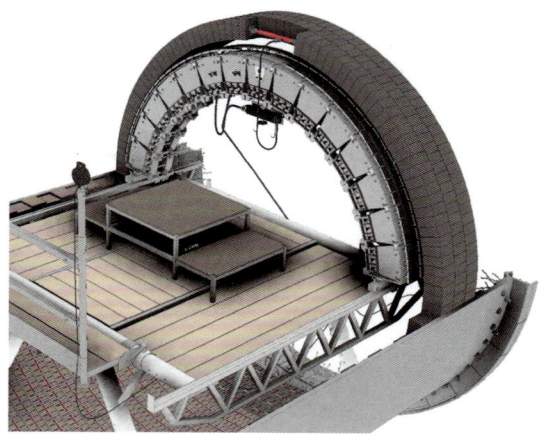

液压千斤顶顶住砖环缺口
Use the hydraulic jack to tighten the brick ring

使用液压千斤顶顶住砖环缺口是砌砖机作业的要求。

不必硬性规定千斤顶的液压压力,压力不应当超过 1/2 火砖冷压强度。在千斤顶挤压期间,要用皮槌敲打火砖以释放砖环中不均匀的静力。

The lining machine requires the use of a hydraulic jack to tighten the brick ring opening. There's no need to specify the pressure of the hydraulic jack, but generally, it should not exceed 1/2 of the brick's cold crushing strength. After tightening the brick ring, use a rubber hammer to knock the ring to distribute the static force within the brick ring.

最后一环的最后一块砖
The last ring and the last brick

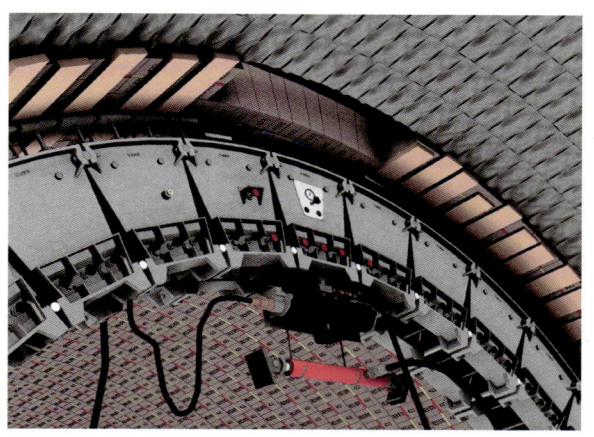

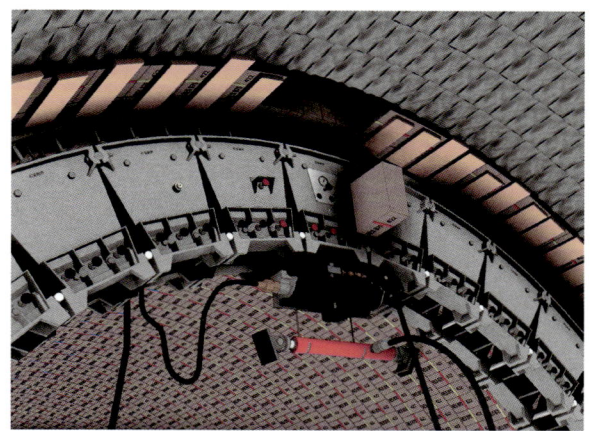

当砌筑最后一环砖时,砖不能自砖环侧面放进环形间隙中,只能如图所示自下而上地放到砖环中,而且单动气缸(第一排)不能直接顶到最后

一环砖上，此时可以用托盘包装上的木板，如图所示来支承最后一环砖。

Often, when lining the final ring of bricks, bricks cannot be put from the side of the ring. Only can be inserted from below, as shown in the illustration. Additionally, the single-action pneumatic cylinders in the first row cannot directly support the bricks of the last ring. In such cases, consider using wooden planks detached from packing pallets to support the last ring bricks, as shown.

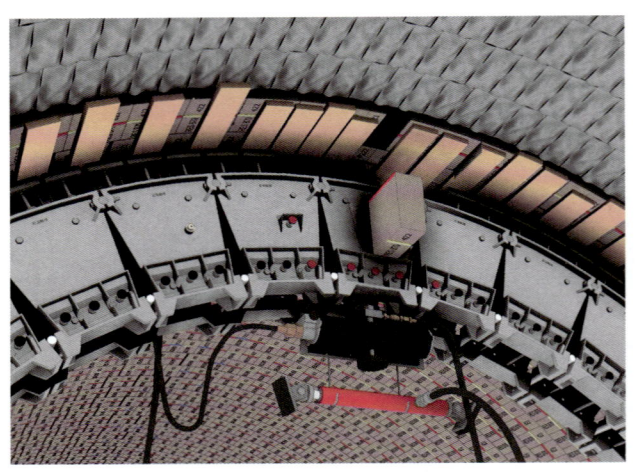
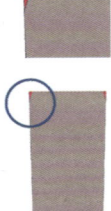

有时，为了把最后一块砖放入最后一环中，需要把砖的大头加工出倒角，不建议这样做但可以接受这种做法。建议加工楔角大的砖。一般地，首先考虑多加一两块钢板来调节最终的间隙。

Sometimes, the cold side of the last brick in the last ring needs to be chamfered to insert in the ring. Although not recommended, it's acceptable. Chamfer the brick with a larger wedge. Generally, first consider not chamfering the brick and use one or two more closure shims to adjust the final gap.

当撤出 DAT 砌砖机时的最终检查
Final check when the DAT is withdrawn from the kiln

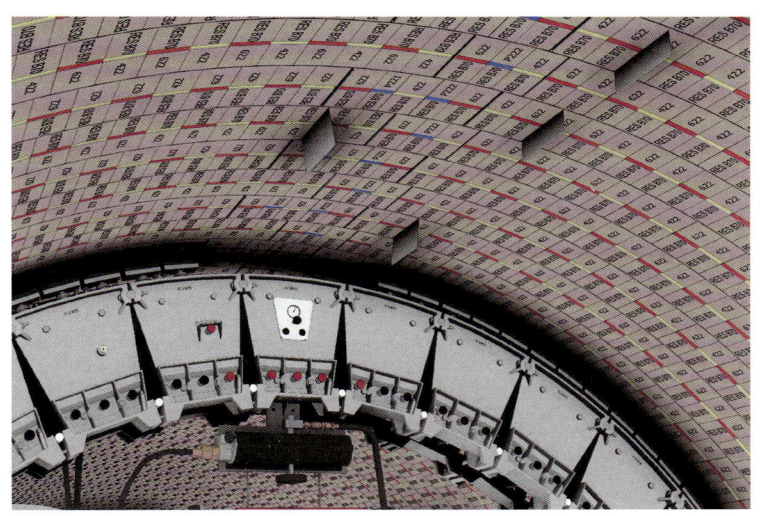

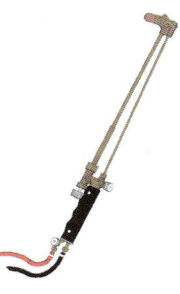

当砌筑完成时，当 DAT 砌砖机自内向外移动时，借助砌砖机平台和照明，用橡皮（木）槌或者用托盘包装拆下的木方，逐环敲打砖环中

的砖来检查，松动的砖环需要补上一到两块锁缝钢板。不能打入凸出在外的锁缝钢板，应当用气割枪或者切割片割除。

When the lining finished, during DAT lining machine withdraw from kiln inside, with the aid of DAT platform and lights illumination, use the rubber(wood) mallet or wood beam detached from packing pallet, knock brick in the rings one by one, to check the tightness of brick rings. Sometimes, one or two more closure shims need to be added to tighten loose rings. Any protruding closure shims that CANNOT be driven in should be cut off by a gas torch or cutting disc.

砖的大小头反了绝对会造成灾难
Bricks placed upside down can lead to absolute disaster

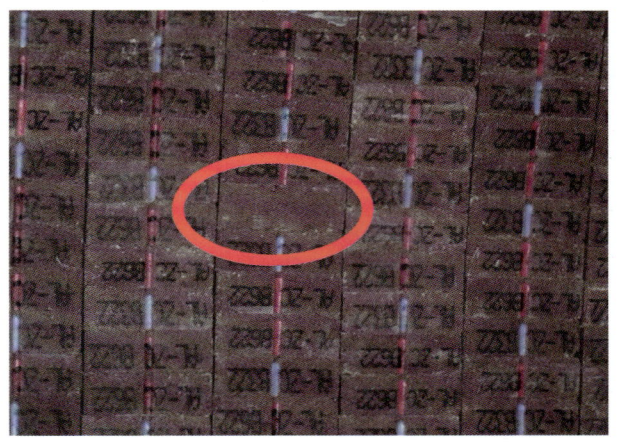

即使经验丰富的筑炉工疲劳时依然会犯这种低级错误。

Even highly experienced bricklayers can make such low-level mistakes when they're exhausted.

砌筑错误一定要查出来并改正！

The lining mistake MUST be identified and corrected!

窑内浇注料施工
Castable installation of rotary kiln

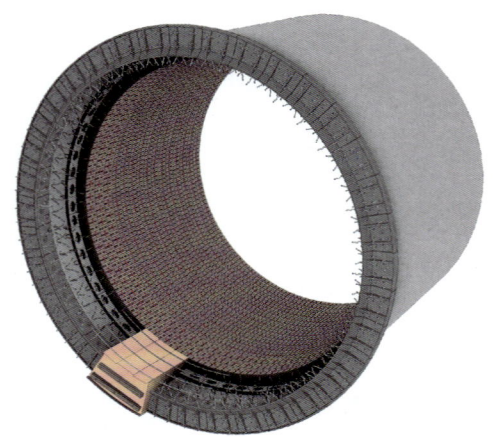

多数情况下，窑内砌砖会和窑口和窑尾锚固件焊接同时进行，待窑内耐火砖砌筑结束后进行窑口和窑尾的浇注料施工。1 块浇注料的弧长和 2 块窑口护铁弧长一致，一般小于 1m。窑口宜选用 310 等级，更加粗的锚固件，且锚固件间距可适当缩小。

In most cases, anchor welding in the outlet and inlet proceeds simultaneously with brick lining inside the kiln. After the brick lining is finished, outlet and inlet castable will be installed. The circumference of one castable segment is normally the same as two nose ring segments, usually less than 1 meter. Grade 310 material is selected for the anchor welded in outlet, which has a larger diameter than normal, and the space between anchors is smaller than normal.

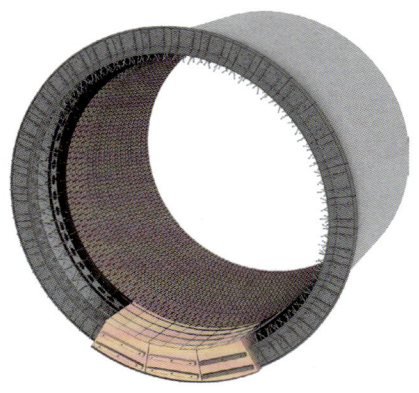

各浇注料扇块之间的膨胀缝间隙为 8~10mm。

The expansion joint between castable segments is between 8-10mm.

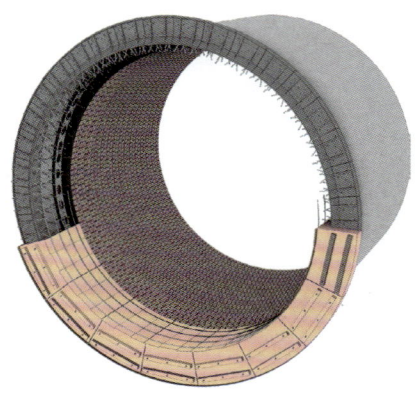

完成一半浇注料的浇筑后,注意养护浇注料,在寒冷地区的冬天还要考虑提高浇注料周边温度来养护。一定要待浇注料硬化后方可转窑,不然会造成安全事故。

After half of the castable is installed, it must cure, especially in the cold winter zones, the freshly installed castable needs to be heated for curing.

Wait until the castable has hardened and gain its necessary strength before rotating the kiln; otherwise, it could lead to safety accidents.

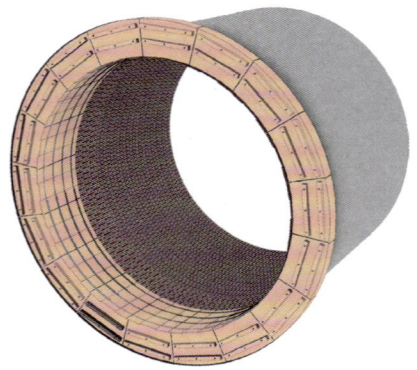

完成另一半浇注料的施工。

Complete the installation of the remaining half of the castable.

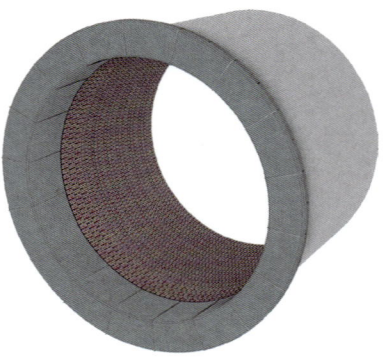

浇注料硬化后,拆除模板。

Once the castable has hardened, remove the molds.

记录和报告

Record and report

应记录基本信息（工作开始、结束日期及时间，砌筑了多少环，砌筑的耐火砖牌号及砌筑位置，砌筑方法等）：

- 每环的理论耐火砖配比及实际火砖配比，每环使用的钢板数量。
- 备注（如窑筒体变形位置，加工砖位置）。
- 照片记录——一张照片胜过千言万语。
- 砌筑最终检查情况记录。

Basic data (start and finish time and date of lining works, rings and brick grades installed, position of brick grades, lining method etc.):

- Comparison of planned and actual mixing ratios per ring, count of key shims per ring.
- Notes (e.g., areas with kiln deformities, brick cutting location).
- Photos — a picture speaks a thousand words.
- Concluding inspection of the installation.

碱性耐火砖的储存
Storage of basic bricks

储存条件：有棚；通风；硬地面；不积水；分区存放不同的砖；先进先出。

取决于包装状态，向材料供应商咨询相关材料在不同气候条件下的最大储存期。

最大堆叠高度 / Maximum pile height	
种类 / category	托盘包装 pallet
镁质砖和非碱性砖 / Magnesia and non basic brick	4
轻质砖 / 隔热砖 /Light weight/insulation brick	2
浇注料，喷涂料，火泥 Castables, gunning mixies, mortars	2

Storage requirements: Roofed; Ventilated; Hard ground; Waterlogging free; Well Partition of different bricks; First in first out.

Depends on the packaging condition, please consult material suppliers for the maximum shelf life of the relevant products in various climate conditions.

碱性耐火砖的水化——未衬砌之前损坏

Hydration of basic bricks—damage before Installation

现象：$MgO+H_2O \xrightarrow[+115\%V]{} Mg(OH)_2$

镁质耐火砖表面有白霜一样的白色物质生成。多数情况下，这类耐火砖依然可用。

镁质耐火砖表面有清晰可见的放射状裂纹。有这种裂纹的砖不可用。

镁质耐火砖表面有鼓包凸起（可用平尺贴靠耐火砖表面来检查）。有这种裂纹的砖不可用。

还可通过锤击敲打辨声来检查。声音响亮无水化，声音闷则有水化。

Phenomena:

White material resembling frost forms on magnesia brick surfaces. In most cases, bricks with this appearance still can be used.

Radial pattern cracks are visible on basic brick surfaces.Brick have such cracks cannot be used.

Surface of basic brick has bulges (can be checked with a straight ruler touching the brick face).Brick with bulge(s) cannot be used.

Also can checked with knocking sound with one metal hammer. A sharp or high-pitched sound indicates no hydration, while a dull or muffled sound indicates hydration.

过高椭圆度造成的耐火砖寿命减少
Short brick service life caused by overdue ovality

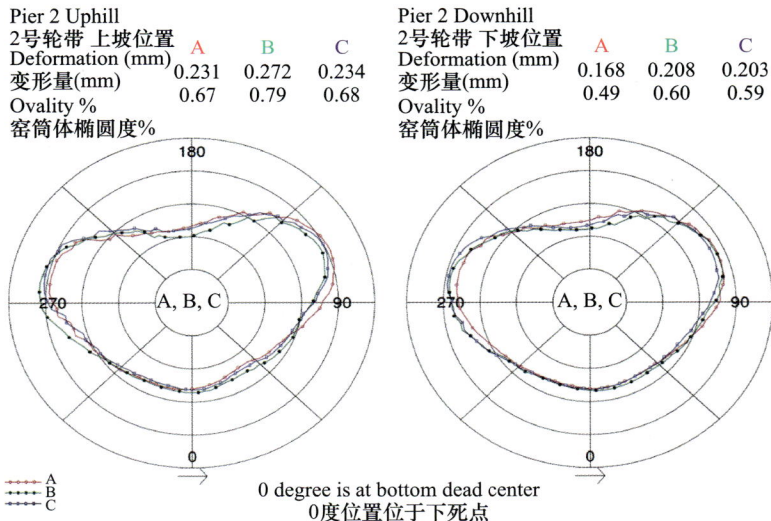

Pier 2 Uphill
2号轮带 上坡位置
Deformation (mm)
变形量(mm)
Ovality %
窑筒体椭圆度%

	A	B	C
Deformation (mm)	0.231	0.272	0.234
Ovality %	0.67	0.79	0.68

Pier 2 Downhill
2号轮带 下坡位置
Deformation (mm)
变形量(mm)
Ovality %
窑筒体椭圆度%

	A	B	C
Deformation (mm)	0.168	0.208	0.203
Ovality %	0.49	0.60	0.59

0 degree is at bottom dead center
0度位置位于下死点

　　定期测量窑筒体椭圆度，不然，没有耐火砖能在高的窑筒体椭圆度下幸存。

依经验，回转窑窑筒体椭圆度小于 [直径（m）/10]% 对于耐火砖是安全的，如窑直径为 5.0m，则小于 0.50%。

轮带间隙会影响回转窑窑筒体椭圆度数值，还有一些其他因素如窑轴线不正、窑筒体有永久性变形、托轮找正不良、回转窑的负荷（窑内填充率和窑的转速）也会影响窑筒体椭圆度。窑筒体锈蚀变薄失去了其应有的刚度，也会导致窑筒体椭圆度数据升高。测量仪器距离轮带侧面的距离也会影响数据的高低。测量点位于焊缝之上也会出现测量值偏高。

To measure kiln kiln shell ovality regularly, otherwise, no brick can survive with high kiln shell ovality.

Rule of thumb, kiln shell ovality figure < [diameter (m)/10]% is safe, e.g. kiln diameter is 5.0m, kiln shell ovality < 0.50% is safe.

The tyre clearance can influence ovality values, other factors such as the alignment of the kiln axis, permanent kiln shell deformations or misalignment of the support rollers, the load of the kiln (filling rate of the kiln and rotation speed) also can influence kiln shell ovality, shell thickness loss due to shell corrosion which make kiln shell lose its due rigidity can increase shell ovality figures, the distance between the tyre and measuring points , also can affect ovality values and measuring points located on welding seams also contribute to higher figures.

窑筒体椭圆度测量的必备条件
Precondition to check kiln shell ovality

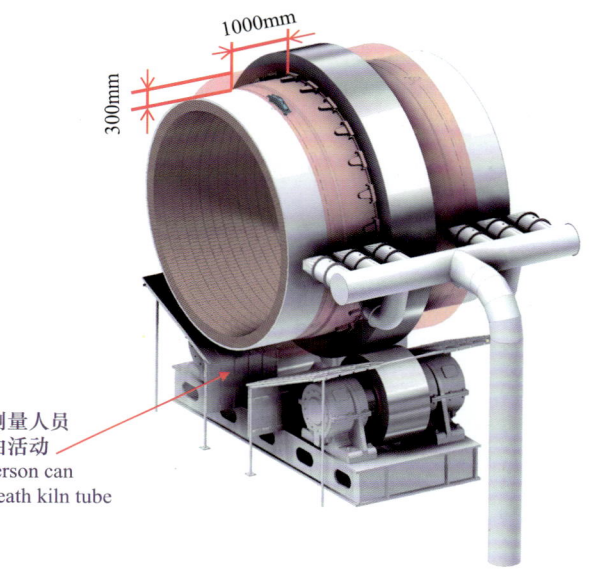

窑筒体正下方测量人员可无阻碍地自由活动
The operating person can move freely beneath kiln tube

- 回转窑满负荷运行。
- 距离轮带侧面 1m 范围、高度 300mm 范围内筒体环向无障碍物。
- Kiln runs at full capacity.
- 1m around fron tyre side face，300mm high circumferential clearance , no obstacles.

定期测量轮带滑移量

Checking tyre creep (migration) regularly

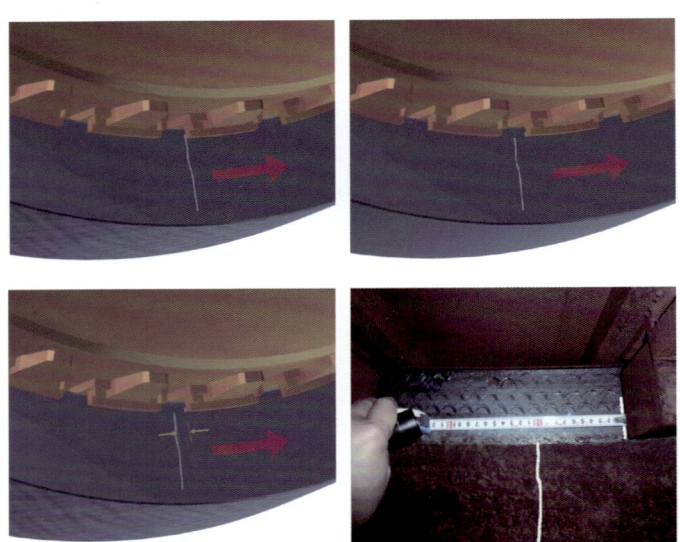

　　回转窑运转时，在轮带侧面使用石笔贴着垫板挡块画出一条线（见插图中的白线），而后待窑转动 5 圈后，测量轮带和挡块端面错开的尺寸，例如 100mm，除以 5，则轮带与筒体相对滑移量为 20mm，测量圈数越多，最终读数误差越小。

　　While rotary kiln is running, hold a chalk lean against side fixing block of pad under tyre, draw one line(as illustration), wait kiln rotates 5 rotations, then measure the displacement between chalk line and side of fixing block of pad, e.g. 100mm measured after 5 rotations，100mm divided by 5, equals 20mm，it's the tyre migration figure. More rotations, less ruler reading errors.

窑筒体铺薄不锈钢板防腐

Sacrifice layer (thin stainless-steel sheet) in case severe kiln shell corrosion

窑筒体铺薄不锈钢板防腐是墨西哥水泥厂发明出来的一种窑筒体防腐方法，是简单可靠的窑筒体防腐方法，近年来，随着越来越多的替代燃料使用，这种方法越来越流行起来。

The sacrificial layer is a method developed by Mexican cement plants to protect the kiln shell from corrosion. It's a simple and reliable way to the kiln shell from corrosion. In recent years, as cement plants have increasingly adopted alternative fuels, this method has gained popularity.

点火升温以及停窑后再次点火升温开窑

Heating-up and restart kiln again after kiln shut down

以下建议针对试车结束的回转窑：

一般地，回转窑点火升温时采用进料室（烟室）热电偶来监控升温状况。点火升温后的初始两三个小时内，系统采用自然通风，进料室温度升温很快，升温速率未必可控，按照规范输入燃料即可。

窑尾进料室温度一旦可控，就要控制燃料输入和系统通风来控制升温速率，对于耐火砖而言，小于50℃/h的升温速率是安全的。升温之前，确认机械，电器及自动化设备运行正常，升温至投料温度将生料投入。按升温操作要求转窑。

经验法则"慢升温，不回头"依然是正确的，点火升温因其他设备故障中断时，如果处理时间不长，不妨继续供给能够让回转窑保温的燃料量。最不好的情况是窑点火升温中断，回转窑冷却，这时，碱性耐火砖上径向砖缝处贴的纸板被烧掉，再次升温往往会在窑中挡砖圈下游出现让窑筒体暴露出来的大缝隙，这种缝隙需要拆砖并用新砖加工来填充缝隙。

The following recommendations are for kilns that have completed commissioning:

Generally, the heating-up temperature of a rotary kiln is monitored by the thermocouple installed in the inlet chamber. In the first 2-3 hours after the kiln is fired, ventilation occurs naturally in the system, causing a rapid increase in inlet chamber temperature. The heating-up rate may become difficult to control during this time, so fuel feeding should be adjusted according to the operating regulations.

Once the inlet chamber temperature is under control, adjust the fuel feeding to the main burner of the kiln to regulate the heating-up rate. For refractory bricks, a heating-up rate of less than 50 ° C/hour is considered safe. Before starting the heating-up process, ensure that all equipment and automation systems are functioning normally. Gradually heat up to the temperature required for feeding raw meal, and then begin rotating the kiln as necessary.

The rule of thumb "heating-up slowly and avoid restart" is correct, if the heating-up interrupted by equipment failure, if the failure could be solved in several hours, keep the kiln warm with suitable fuel feeding. The worst case is heating-up forced by other failure, and the kiln must be cooled down, if the cardboard on brick radial joint burnt out, during subsequent kiln restart, one ring gap would be formed in downstream of middle brick retaining ring, which could make kiln shell exposed, and such ring gap need to remove one or two ring installed bricks, and install new bricks to fill in the gap.

回转窑中的主导损耗因素

Primary wear mechanisms in cement rotary kilns

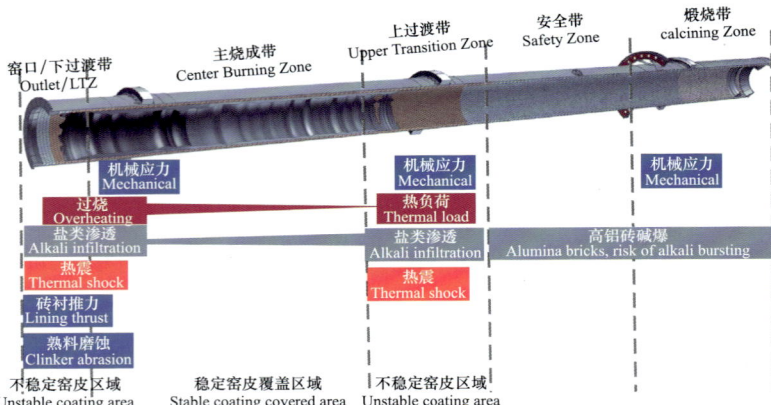

上图概括了回转窑内各区域的耐火砖主导的损坏原因

The above illustration summarizes the predominant wear mechanisms in different areas of cement rotary kilns.

窑皮就是冷却了的熟料液相,稳定窑皮位置就是烧成带的位置,近似的窑喷煤管火焰长度。

Coating is the cooled clinker melt, stable coating position is the burning zone area, approximately the flame length of kiln main burner.

水泥回转窑机械状况常识
The common knowledge of cement rotary kiln mechanical condition

一条典型的三挡支承回转窑

A typical cement rotary with 3 tyres

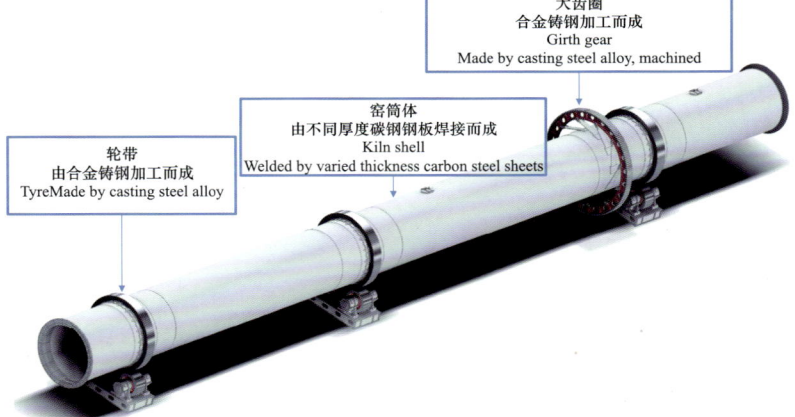

轮带
由合金铸钢加工而成
TyreMade by casting steel alloy

窑筒体
由不同厚度碳钢钢板焊接而成
Kiln shell
Welded by varied thickness carbon steel sheets

大齿圈
合金铸钢加工而成
Girth gear
Made by casting steel alloy, machined

窑筒体：回转的物料热处理工作容器。

大齿圈：驱（拉）动工作容器（圆筒）转动。

轮带和托轮：支承窑筒体/耐火砖/处理物料质量保持窑筒体成为圆形形状。

Kiln shell: rotating material pyro-processing container.

Girth gear: driving the container(tube) rotation.

Tyres and supporting roller sets: support weight of kiln tube/refractory/material keep the kiln tube maintain round shape.

我们期望的机械状态

The condition we expect

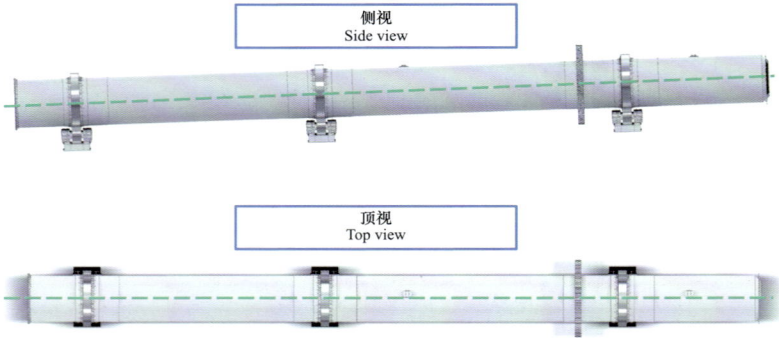

实际上

Actual

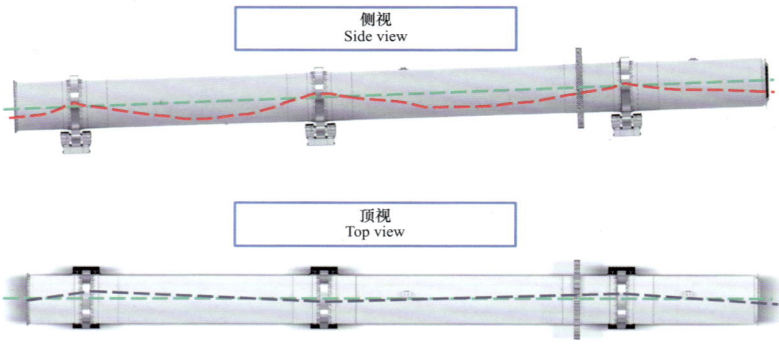

当筒体转动时,扭力是始终存在的
Torsion always exist when the tube rotating

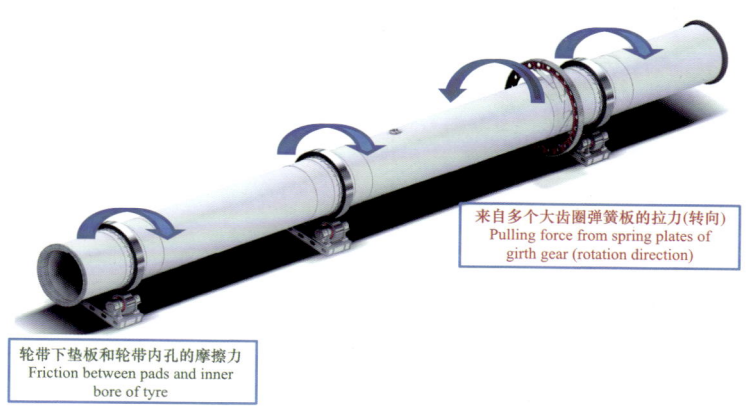

来自多个大齿圈弹簧板的拉力(转向)
Pulling force from spring plates of girth gear (rotation direction)

轮带下垫板和轮带内孔的摩擦力
Friction between pads and inner bore of tyre

轮带的摆动也会施加不期望的应力
Wobbling of tyre also can exert unexpected stress on shell

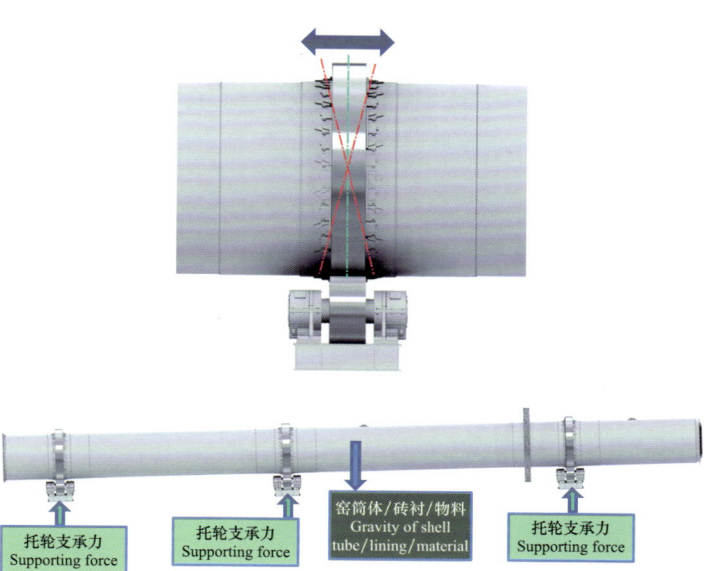

窑筒体/砖衬/物料
Gravity of shell tube/lining/material

托轮支承力
Supporting force

托轮支承力
Supporting force

托轮支承力
Supporting force

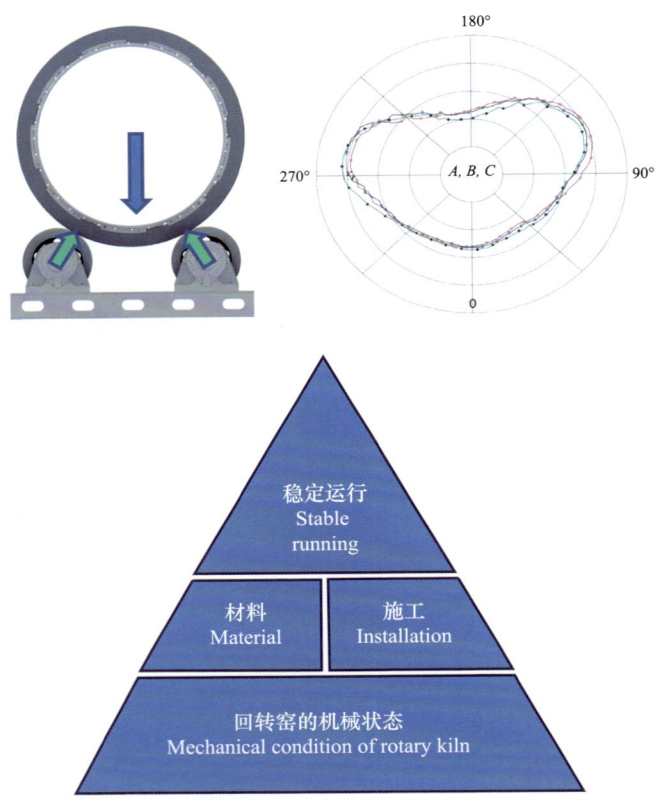

回转窑的机械状态对耐火砖的使用寿命影响很大，窑筒体椭圆度过大，窑轴线找正不良，窑筒体锈蚀失去刚度等情况，都会对窑内耐火砖产生不利的影响，这是影响窑内耐火砖寿命的最底层基础。

The mechanical condition of a rotary kiln significantly impacts the service life of the bricks lined inside rotary kiln, overdue kiln shell ovality, misalignment of kiln axis, shell rigidness losing caused by shell corrosion etc. will create adverse effects to bricks lined inside kiln, it's the lowest foundation of pyramid.

未来可期
The future is promising

随着技术的不断进步，测控元件不断进步，机器人应用越来越广泛，人工智能不断进步，可以预见，未来结合了筑炉经验和技艺的人工智能筑炉机器人会出现，让工人们从繁重的筑炉工作中解脱出来。

As technology advances, sensors and control elements continue to improve, and robots are being used more widely, AI technology continues to advance. It is anticipated that an automated lining robot will emerge, which would combines masonry expertise and craftsmanship. This development could free bricklayers from the heavy labor work of kiln lining.

参考文献
Reference

[1] JOSEF NIEVOLL.Correct installation of rotary kiln bricks[Z].

[2] ROLAND KRISCHANITZ.Wear mechanism[Z].

[3] ECKHARD HOBRECHT.REPORT No.41 installation of refractory materials in cement rotary kilns and corresponding aggregates[Z].